Modern Mathematics - A Handbook of Definitions and Theorems

Luther Rinehart

In memory of Paul J. Sally Jr.

iv

Contents

Introduction

This work is an outline of the main definitions and theorems of modern mathematics. It is intended as a reference for non-mathematicians whose work requires knowledge and use of advanced mathematical techniques, as well as a guide for those intimidated by the proliferation of nomenclature and notation. The material is arranged pedagogically, beginning with first principles and building new definitions on top of prior ones. This is not a math textbook. It is no substitute for a disciplined and proof-based study of mathematics.

Modern mathematics is characterized by rigorous proof of propositions within a single formal framework, in which all concepts are explicitly defined in terms of a few first principles. It is not the abstract manipulation of numbers and symbols, but the concrete manipulation of well-defined objects of rational thought.

This handbook attempts to present mathematics through a unified perspective, showing the connections and theoretical foundations of the various concepts in the most general terms possible. An effort is made to introduce no more terminology and concepts than are necessary, and to minimize redundancy. From this perspective, many of the topics usually thought to comprise the content of mathematics are merely special cases or applications of the very general concepts presented here, and so are not treated individually.

Building on the foundation of set theory, the two main themes of mathematics are algebra and topology. The former is the study of 'operational' relationships, the latter is the study of 'spatial' relationships. With these core principles in place, the chapter 'Analysis et al.' combines the two in various interesting ways.

It goes without saying that the list of concepts and theorems presented here is incomplete. First, no single volume can address all possible topics of interest, and some selection is necessary. This selection was done based on the author's bias concerning what is useful and interesting. Second, modern mathematics is an active field of research. This handbook will inevitably become less comprehensive as new concepts are formulated and new results are proved.

Chapter 1

Set Theory

The foundation of mathematics is set theory. All mathematical entities are objects known as sets, which are interpreted as collections of other objects, themselves sets. There is only one primitive relation which exists between sets, called $\in$, and all mathematical statements can be reduced to statements about $\in$. For two sets x, y, the relation is written $x \in y$, and is interpreted to mean that x is an element of y. The nature of $\in$ is characterized by a collection of axioms which formalize what it means for a set to be a collection of elements, and which constitute the first principles of mathematics.

Axiom 1. (Equality) The condition "$\forall z, \ x \in z \Leftrightarrow y \in z$" holds iff "$\forall z, \ z \in x \Leftrightarrow z \in y$" also holds.

This means x and y are elements of the same sets iff they have the same elements.

Definition 1. x is *equal* to y iff the conditions in the axiom of equality hold. That is,
$$x = y \text{ iff } \forall z, \ z \in x \Leftrightarrow z \in y$$

Axiom 2. (Specification) For any set x and any definable property P, there exists a set y such that $z \in y$ iff $z \in x$ and $P(z)$ holds.

Such a set is notated as $y = \{z \in x \mid P(z)\}$.

Definition 2. The *empty set* $\varnothing$ is the set having no elements. $\forall x, x \notin \varnothing$.

It can be defined by $\varnothing = \{z \in x \mid z \notin x\}$.

Axiom 3. (Pairing) $\forall x, y$ there exists a set having exactly x and y as elements.

This can also be used to form the set having exactly x as an element, which, by the axiom of equality, is equal to the pairing of x with itself.

Axiom 4. (Union) For any set x, there exists a set $\bigcup x$ such that $z \in \bigcup x$ iff $z \in y$ for some $y \in x$.

$\bigcup x$ is said to be the union of the elements of x.

Definition 3. Let $x = \{y, z\}$, the pairing of y and z. Define the *union* of y and z as

$$y \cup z \equiv \bigcup x$$

The axioms of pairing and union can now be used to build larger sets out of existing sets.

Definition 4. The *intersection* of x and y is

$$x \cap y \equiv \{z \in x \cup y \mid z \in x \text{ and } z \in y\}$$

and likewise

$$\bigcap x \equiv \{z \in \bigcup x \mid z \in y \; \forall y \in x\}$$

Definition 5. x is a *subset* of y, $x \subseteq y$, iff $\forall z \in x, z \in y$.

Facts:

$x \cup \varnothing = x$
$x \cup y = y \cup x$
$x \cup (y \cup z) = (x \cup y) \cup z$
$x \cup x = x$

$x \cap \varnothing = \varnothing$
$x \cap y = y \cap x$
$x \cap (y \cap z) = (x \cap y) \cap z$
$x \cap x = x$

$x \cap (y \cup z) = (x \cap y) \cup (x \cap z)$
$x \cup (y \cap z) = (x \cup y) \cap (x \cup z)$

$x \subseteq y$ iff $x \cup y = y$ iff $x \cap y = x$
If $x \subseteq y$ then $\bigcup x \subseteq \bigcup y$
If $x \subseteq y$ then $\bigcap y \subseteq \bigcap x$
If $x \subseteq y$ and $z \subseteq w$ then $(x \cup z) \subseteq (y \cup w)$
If $x \subseteq y$ and $z \subseteq w$ then $(x \cap z) \subseteq (y \cap w)$
If $x \subseteq y$ and $y \subseteq z$ then $x \subseteq z$

Definition 6. $A \backslash B \equiv \{C \in A \mid C \notin B\}$. Let $A \subseteq X$. The *compliment* of A in X is $A^c \equiv X \backslash A$.

Facts:

$A^{cc} = A$
$\varnothing^c = X$
$A \cap A^c = \varnothing$
$A \cup A^c = X$
$A \subseteq B$ iff $B^c \subseteq A^c$
$(A \cup B)^c = A^c \cap B^c$
$(A \cap B)^c = A^c \cup B^c$

Axiom 5. (Powers) For any set x, there exists a set $\mathcal{P}(x)$ such that $z \in \mathcal{P}(x)$ iff $z \subseteq x$.

Definition 7. The *Cartesian product* of sets A and B is

$$A \times B \equiv \left\{ \{\{a, b\}, \{a\}\} \in \mathcal{P}\mathcal{P}(A \cup B) \mid a \in A \text{ and } b \in B \right\}$$

Elements of a Cartesian product are called *ordered pairs*, and are notated

$$\{\{a,b\},\{a\}\} = (a,b)$$

Facts:

$(A \cup B) \times C = (A \times C) \cup (B \times C)$
$(A \cap B) \times (C \cap D) = (A \times C) \cap (B \times D)$
$A \times \varnothing = \varnothing \times A = \varnothing$
If $A \subseteq C$ and $B \subseteq D$ then $(A \times B) \subseteq (C \times D)$

Definition 8. A *relation* R between A and B is a subset of $A \times B$. If $(a,b) \in R$ then write aRb.

Definition 9. A relation $<$ between X and itself is an *order* iff $\forall x, y, z \in X$,
1. Either $x < y$ or $y < x$ or $x = y$, but not more than one of these.
2. If $x < y$ and $y < z$ then $x < z$

Definition 10. In an ordered set, $x \leq y$ iff $x < y$ or $x = y$. An *interval* is a subset of one of the following types:
$(a,b) \equiv \{x \mid a < x < b\}$
$[a,b] \equiv \{x \mid a \leq x \leq b\}$
$(a,b] \equiv \{x \mid a < x \leq b\}$
$[a,b) \equiv \{x \mid a \leq x < b\}$

Definition 11. Let X have an order, and let $A \subseteq X$. x is an *upper bound* of A iff $\forall a \in A$, $a < x$. x is a *lower bound* of A iff $\forall a \in A$, $x < a$.

Definition 12. An ordered set satisfies *least upper bound* iff every subset bounded above has a least upper bound, denoted $\sup A$. An ordered set satisfies *greatest lower bound* iff every subset bounded below has a greatest lower bound, denoted $\inf A$.

Theorem 1. An order satisfies least upper bound iff it satisfies greatest lower bound.

Definition 13. An order is *dense* iff $\forall x, \forall y < x$, $\exists z$ such that $y < z < x$.

Definition 14. A relation $\sim$ between X and itself is an *equivalence* iff $\forall x, y, z \in X$,
1. $x \sim x$
2. If $x \sim y$ then $y \sim x$
3. If $x \sim y$ and $y \sim z$ then $x \sim z$

The *equivalence class* of $x \in X$ is $\{y \in X \mid x \sim y\}$.

Definition 15. A *partition* of X is $P \subseteq \mathcal{P}(X)$ satisfying
1. $\forall p, p' \in P, \; p \cap p' = \varnothing$
2. $\bigcup P = X$
3. $\varnothing \notin P$

Theorem 2. A set $P \subseteq \mathcal{P}(X)$ is a partition iff it is the set of equivalence classes of an equivalence relation on X.

Definition 16. A relation $f \subseteq A \times B$ is a *function* $f \colon A \to B$ iff
1. $\forall x \in A \; \exists y \in B$ such that $(x, y) \in f$
2. If $(x, y) \in f$ and $(x, z) \in f$ then $y = z$

If $(x, y) \in f$, write $f(x) = y$. The set of functions from A to B is denoted B^A. If $C \subseteq A$ and $D \subseteq B$, then define

$$f(C) \equiv \{f(c) \mid c \in C\} \qquad f^{-1}(D) \equiv \{a \mid f(a) \in D\}$$

Facts:

$A^X \cap B^X = (A \cap B)^X$
$A^X \cup B^X \subseteq (A \cup B)^X$
If $A \subseteq B$ then $A^X \subseteq B^X$
$X^{A \times B} = (X^A)^B$
$(A \times B)^X = A^X \times B^X$
$A^X = A^B \times A^{B^c}$
$f(C \cup D) = f(C) \cup f(D)$
$f(C \cap D) \subseteq f(C) \cap f(D)$
If $C \subseteq D$ then $f(C) \subseteq f(D)$

Definition 17. Let $f\colon A \to B$ and $g\colon B \to C$. Their *composition* is the function $g \circ f\colon A \to C$ such that $g \circ f(x) = g(f(x))$.

Definition 18. Let $f\colon A \to B$ and $g\colon C \to D$. Define the function $f \times g\colon A \times C \to B \times D$ as $f \times g\,(x, y) = (f(x), g(y))$.

Definition 19. A function $f\colon A \to B$ is *constant* iff $\forall x, y \in A$, $f(x) = f(y)$.

Definition 20. A function $f\colon A \to A$ is *identity* iff $\forall x \in A$, $f(x) = x$.

Definition 21. A function $f\colon A \to B$ is *injective* iff

$$\text{If } f(x) = z \text{ and } f(y) = z \text{ then } x = y$$

It is *surjective* iff

$$\forall y \in B \ \exists x \in A \text{ such that } f(x) = y$$

It is a *bijection* iff it is injective and surjective.

Definition 22. For any function $f\colon A \to B$, a function $g\colon B \to A$ is its *inverse* iff $\forall a \in A$, $g(f(a)) = a$, and $\forall b \in B$, $f(g(b)) = b$.

Theorem 3. A function is a bijection iff it has an inverse.

Definition 23. Two sets have the same *cardinality* iff there exists a bijection between them.

Theorem 4. On any set, the property of having the same cardinality is an equivalence relation.

Theorem 5. The following are equivalent:
1. A has the same cardinality as a subset of B
2. $\exists$ injection $A \to B$
3. $\exists$ surjection $B \to A$

Theorem 6. The following are equivalent:
1. A has the same cardinality as B
2. $\exists$ injections $A \to B$ and $B \to A$
3. $\exists$ surjections $A \to B$ and $B \to A$

Definition 24. An *indexed family* of sets $\{x_i\}_{i \in I}$ is a function $f \colon I \to X$. The set I is the indexing set, and the elements of the family are $x_i = f(i)$.

Definition 25. The *Cartesian product* $\prod_{i \in I} x_i$ of an indexed family of sets is the set of functions $g \colon I \to \bigcup_{i \in I} x_i$ satisfying $\forall i,\ g(i) \in x_i$.

Axiom 6. (Choice) Every Cartesian product of nonempty sets is nonempty.

Definition 26. A set x is *infinite* iff $\exists y \subsetneq x$ such that x and y have the same cardinality. A set is *finite* if it is not infinite.

Theorem 7. Every subset of a finite set is finite.

Theorem 8. A finite union of finite sets is finite. A finite product of finite sets is finite.

Definition 27. For any set x, the *successor* of x is the set $x^+ \equiv x \cup \{x\}$.

Axiom 7. (Infinity) For any set x there exists a set y such that $x \in y$ and $\forall z \in y,\ z^+ \in y$ also.

This axiom completes the list of axioms of set theory. In summary, they are
1. Equality
2. Specification
3. Pairing
4. Union
5. Powers
6. Choice
7. Infinity

Definition 28. A set A is *successor* iff $\varnothing \in A$ and $\forall x \in A,\ x^+ \in A$.

Theorem 9. Every successor set is infinite.

Definition 29. Let A be successor. Define the *natural numbers*

$$\mathbb{N} \equiv \bigcap \{x \in \mathcal{P}(A) \mid x \text{ is successor}\}$$

This is well defined since

Theorem 10. $\forall A'$ successor,

$$\bigcap \{x \in \mathcal{P}(A') \mid x \text{ is successor}\} = \mathbb{N}$$

Theorem 11. $n \in \mathbb{N}$ iff $n \in A$ $\forall A$ successor.

Definition 30. $\varnothing \in \mathbb{N}$. Define $0 \equiv \varnothing$. Define $1 \equiv 0^+$ etc.

Theorem 12. If $B \subseteq \mathbb{N}$ is successor, then $B = \mathbb{N}$.

Theorem 13. Let $g \colon X \to X$, $x \in X$. $\exists$ a unique $f \colon \mathbb{N} \to X$ such that $f(0) = x$ and $\forall n$, $f(n^+) = g(f(n))$.

Definition 31. Let $g \colon X \to X$. The *exponentiation* of g is the function $\mathbb{N} \to X^X$ such that $g^0(x) = x$ and $g^{n^+}(x) = g(g^n(x))$.

Definition 32. $S \colon \mathbb{N} \to \mathbb{N}$ $S(n) \equiv n^+$

Facts:

$\forall n$, $0 \neq S(n)$
If $S(n) = S(m)$ then $n = m$

Definition 33.

$$m + n \equiv S^m(S^n(0))$$

$$m \cdot n \equiv (S^m)^n(0)$$

Theorem 14. $\forall g \colon X \to X$, $g^m \circ g^m = g^{m+n}$ and $(g^m)^n = g^{m \cdot n}$.

Definition 34. $m < n$ iff $m \in n$.

Theorem 15. $<$ is an order.

Facts:

$a + b = b + a$
$(a + b) + c = a + (b + c)$
$a + 0 = a$
$a \cdot b = b \cdot a$
$(a \cdot b) \cdot c = a \cdot (b \cdot c)$
$a \cdot 1 = a$
$a \cdot 0 = 0$
$a \cdot (b + c) = (a \cdot b) + (a \cdot c)$
$a = b$ iff $a + c = b + c$
$a < b$ iff $a + c < b + c$
$a = b$ iff $a \cdot c = b \cdot c$
$a < b$ iff $a \cdot c < b \cdot c$

Theorem 16. X is finite iff there is a unique natural number $\#X$ with the same cardinality as X.

Definition 35. A set is *countable* iff it is finite, or has the same cardinality as $\mathbb{N}$.

Theorem 17. Every subset of a countable set is countable.

Theorem 18. A countable union of countable sets is countable. A finite product of countable sets is countable.

Theorem 19. $\mathcal{P}(\mathbb{N})$ is uncountable.

Facts:

If $n \in \mathbb{N}$ then $\#n = n$
If $A \subseteq B$ then $\#A \le \#B$
$\#(A \cup B) = \#A + \#B - \#(A \cap B)$
$\#(A \setminus B) = \#A - \#(A \cap B)$
$\#(A \times B) = \#A \cdot \#B$
$\#(A^B) = \#A^{\#B}$
$\#\mathcal{P}(A) = 2^{\#A}$
$\#$ bijections $A \to A = (\#A)!$

Chapter 2

Algebra

2.1 Group Theory

Definition 36. A *group* is a set G with a function $*\colon G \times G \to G$, written as $*(g, h) \equiv gh$, satisfying
1. (associative) $(gh)k = g(hk)$
2. (identity) $\exists e \in G$ such that $\forall g \in G,\ eg = ge = g$
3. (inverse) $\forall g \in G,\ \exists g^{-1} \in G$ such that $gg^{-1} = g^{-1}g = e$

Theorem 20. If $\exists h, g \in G$ such that $hg = g$, then $h = e$. If $\exists h, g \in G$ such that $hg = e$, then $h = g^{-1}$.

Theorem 21. If $gh = gk$ then $h = k$.

Theorem 22. $(gh)^{-1} = h^{-1}g^{-1}$

Theorem 23. $(g^{-1})^{-1} = g$

Definition 37. A group G is *abelian* iff $\forall g, h \in G,\ gh = hg$.

Definition 38. The *order* of a group is G is its cardinality $\#G$.

Definition 39. A subset $H \subseteq G$ is a *subgroup* of G, denoted $H \leq G$, iff it is itself a group under $*$.

Theorem 24. $H \subseteq G$ is a subgroup iff
1. $g, h \in H \Rightarrow gh \in H$
2. $e \in H$
3. $g \in H \Rightarrow g^{-1} \in H$

Theorem 25. For any two subgroups $H, K \leq G$, $H \cap K$ is a subgroup.

Definition 40. Let G, G' be groups. A function $\phi \colon G \to G'$ is a *homomorphism* iff $\forall g, h \in G$, $\phi(gh) = \phi(g)\phi(h)$.

Theorem 26. If ϕ is a homomorphism, then:
$\phi(e) = e$
$\phi(g^{-1}) = \phi(g)^{-1}$
$H \leq G \Rightarrow \phi(H) \leq G'$
$K \leq G' \Rightarrow \phi^{-1}(K) \leq G$

Definition 41. Let ϕ be a homomorphism. $\ker \phi \equiv \{g \in G \mid \phi(g) = e\}$.

Theorem 27. $\ker \phi \leq G$

Definition 42. An *isomorphism* is a homomorphism that is bijective. Two groups are *isomorphic*, denoted $G \simeq G'$, iff there exists an isomorphism between them.

Theorem 28. On any set of groups, isomorphism is an equivalence relation.

Theorem 29. If ϕ is an isomorphism, then so is ϕ^{-1}.

Theorem 30. A homomorphism ϕ is injective iff $\ker \phi = \{e\}$.

Theorem 31. $\forall g \in G$, $i_g \colon G \to G$ such that $i_g(h) = ghg^{-1}$ is an isomorphism.

Theorem 32. For any groups G, H, $G \times H$ is a group with $(g, h)(g', h') \equiv (gg', hh')$.

Theorem 33. $G \times H \simeq H \times G$

Theorem 34. If G and H are abelian, then so is $G \times H$.

Theorem 35. Let $H, K \leq G$. $G \simeq H \times K$ iff
1. $\forall g \in G$, $\exists (h, k) \in H \times K$ such that $g = hk$
2. $\forall (h, k) \in H \times K$, $hk = kh$
3. $H \cap K = \{e\}$

Definition 43. Let $H \leq G$, $g \in G$. $gH \equiv \{gh \mid h \in H\}$ and $Hg \equiv \{hg \mid h \in H\}$ are *cosets* of H.

Theorem 36. $\{gH \mid g \in G\}$ and $\{Hg \mid g \in G\}$ are partitions of G.

Theorem 37. If G is finite and $H \leq G$, then $\exists m \in \mathbb{N}$ $\#G = m\#H$.

Theorem 38. $\forall g \in G$, $\#gH = \#Hg = \#H$.

Theorem 39. $\#\{gH \mid g \in G\} = \#\{Hg \mid g \in G\} = \#G/\#H$.

Definition 44. For any group, $(G : H) \equiv \#\{gH \mid g \in G\}$

Theorem 40. If $K \leq H \leq G$, $(G : K) = (G : H)(H : K)$.

Definition 45. A subgroup $N \leq G$ is *normal*, denoted $N \trianglelefteq G$, iff $\forall g \in G$, $gN = Ng$.

Theorem 41. If ϕ is a homomorphism, $\ker \phi$ is normal.

Theorem 42. Let $N \trianglelefteq G$. The *factor group* $G/N \equiv \{gN \mid g \in G\}$ is a group under $(gN)(hN) \equiv (gh)N$.

Theorem 43. N is the identity in G/N.

Theorem 44. $G/\ker \phi \simeq \phi(G)$

Theorem 45. Every subgroup of an abelian group is normal. Every factor group of an abelian group is abelian.

Theorem 46. If $H \trianglelefteq G$ and $(G : H) = m$, then $\forall g \in G$, $g^m \in H$.

Theorem 47. If $H, K \trianglelefteq G$, then $H \cap K \trianglelefteq G$.

Theorem 48. $H, K \trianglelefteq H \times K$, and $(H \times K)/K \simeq H$.

Theorem 49. Let $\phi \colon G \to G'$ be a homomorphism.
If $N \trianglelefteq G$ then $\phi(N) \trianglelefteq \phi(G)$.
If $N' \trianglelefteq \phi(G)$ then $\phi^{-1}(N') \trianglelefteq G$.

Theorem 50. If $N \leq H \leq G$ and $N \trianglelefteq G$, then $N \trianglelefteq H$.

Theorem 51. Let $H \leq G$, $N \trianglelefteq G$. Let $HN \equiv \{hn \mid h \in H, \ n \in N\}$.
Then $H, N \leq HN \leq G$.

Theorem 52. If $H \leq G$ and $N_1 \trianglelefteq N_2 \leq G$, then $(H \cap N_1) \trianglelefteq (H \cap N_2)$.
If $H \trianglelefteq G$ and $N_1 \trianglelefteq N_2 \leq G$, then $HN_1 \trianglelefteq HN_2$.

Theorem 53. $(G \times K) \simeq (H \times K)$ iff $G \simeq H$.

Theorem 54.
$$\frac{G}{N} \times \frac{H}{M} \simeq \frac{G \times H}{N \times M}$$

Theorem 55. Let $K \trianglelefteq H \trianglelefteq G$.
$$\frac{G/K}{H/K} \simeq \frac{G}{H}$$

Theorem 56.
$$\frac{HN}{N} \simeq \frac{H}{H \cap N}$$

Theorem 57. Let $N \trianglelefteq H, K \leq G$.
$H/N \simeq K/N$ iff $H \simeq K$
$H \leq K$ iff $(H/N) \leq (K/N)$
$(H \cap K)/N = (H/N) \cap (K/N)$
$H \trianglelefteq G$ iff $H/N \trianglelefteq G/N$
$(HK)/N = (H/N)(K/N)$

Definition 46. A group G is *simple* iff its only normal subgroups are $\{e\}$ and G.

Theorem 58. Let $N \trianglelefteq G$. G/N is simple iff there is no $M \triangleleft G$ with $N < M < G$.

Theorem 59. Let $H \leq G$, let $N(H) \equiv \{g \in G \mid gHg^{-1} = H\}$. Then $H \trianglelefteq N(H) \leq G$.

Definition 47. Elements $h, k \in G$ are *conjugate* iff $\exists g \in G$ such that $ghg^{-1} = k$. Subgroups $H, K \leq G$ are *conjugate* iff $\exists g \in G$ such that $gHg^{-1} = K$.

Theorem 60. Conjugacy is an equivalence relation on G, and an equivalence relation on subgroups. The equivalence classes on G are called *conjugacy classes*.

Theorem 61. If $H, K \leq G$ are conjugate, then $H \simeq K$.

Definition 48. The *center* of G is $Z(G) \equiv \{g \in G \mid hgh^{-1} = g \ \forall h \in G\}$.

Theorem 62. $Z(G)$ is an abelian normal subgroup of G.

Theorem 63. Let G be a group and $A \subseteq G$. There is a unique subgroup $\langle A \rangle \leq G$, *generated* by A, such that $A \subseteq \langle A \rangle$ and $\forall H \leq G$, if $A \subseteq H$ then $\langle A \rangle \leq H$.

Theorem 64. Let $C(G) \equiv \langle \{ghg^{-1}h^{-1} \mid g, h \in G\} \rangle$. $C(G) \trianglelefteq G$, $G/C(G)$ is abelian, and $\forall N \trianglelefteq G$, G/N is abelian iff $C(G) \leq N$.

Definition 49. Let $\sim$ be the equivalence relation on $\mathbb{N} \times \mathbb{N}$ such that $(n, m) \sim (n', m')$ iff $n + m' = m + n'$. Let $\mathbb{Z}$, the *integers*, be the set of equivalence classes under $\sim$. Define $+: \mathbb{Z} \times \mathbb{Z} \to \mathbb{Z}$ as $(n, m) + (n', m') \equiv (n + n', m + m')$.

Theorem 65. $\mathbb{Z}$ is an abelian group under $+$. Letting the equivalence class of $(n, 0)$ be denoted by n, the identity is 0, and the inverse of n is $-n \equiv (0, n)$.

Theorem 66. $\mathbb{Z}$ has a natural order given by $(n, m) < (n', m')$ iff $n + m' < m + n'$.

Theorem 67. For $n \in \mathbb{N}$, let $n\mathbb{Z} \equiv \langle n \rangle = \{nm \mid m \in \mathbb{Z}\} \leq \mathbb{Z}$. $n\mathbb{Z} \trianglelefteq \mathbb{Z}$.

Definition 50. $\mathbb{Z}_n \equiv \mathbb{Z}/n\mathbb{Z}$

Theorem 68. $\#\mathbb{Z}_n = n$

Definition 51. A group G is *cyclic* iff $\exists a \in G$ such that $G = \langle a \rangle$.

Theorem 69. $\langle a \rangle = \{a^n \mid n \in \mathbb{Z}\}$

Theorem 70. $\mathbb{Z}_n$ is cyclic.

Theorem 71. If G is cyclic then all subgroups of G are cyclic.

Theorem 72. All cyclic groups are abelian.

Theorem 73. If G is cyclic then all factor groups of G are cyclic.

Theorem 74. If $\#G$ is prime, then G is cyclic.

Theorem 75. If G is cyclic and $\#G = n$, then $G \simeq \mathbb{Z}_n$.

Theorem 76. If G is cyclic and infinite, then $G \simeq \mathbb{Z}$.

Theorem 77. Let G be cyclic of order n. Let $G = \langle a \rangle$.
$\#\langle a^s \rangle = n/\gcd(n, s)$, where $\gcd(n, s)$ denotes the *greatest common divisor* of n and s.
$a^k = e$ iff $\exists m \in \mathbb{Z}\ k = mn$.
$\langle a^s \rangle = G$ iff $\gcd(n, s) = 1$.
$\forall k$ such that $\exists m \in \mathbb{N}\ n = mk$, there is a unique subgroup of G of order k.
$\langle a^s \rangle = \langle a^r \rangle$ iff $\gcd(n, s) = \gcd(n, r)$.

Theorem 78. $\mathbb{Z}_n \times \mathbb{Z}_m \simeq \mathbb{Z}_{nm}$ iff $\gcd(n, m) = 1$.

Theorem 79. Let $a \in G$, $b \in H$. If $\#\langle a \rangle = n$ and $\#\langle b \rangle = m$, then in $G \times H$, $\#\langle (a, b) \rangle = \mathrm{lcm}(n, m)$, where lcm denotes the *least common multiple* of n and m.

Theorem 80. In any group G, $g \in G$,

$$\langle g^n \rangle \cap \langle g^m \rangle = \langle g^{\mathrm{lcm}(n,m)} \rangle$$

$$\langle g^n, g^m \rangle = \langle g^{\gcd(n,m)} \rangle$$

Theorem 81. $nm = \text{lcm}(n, m) \gcd(n, m)$

Theorem 82. Let G be finite abelian. For each prime $p \in \mathbb{N}$ and each power $r \in \mathbb{N}$, there is a unique $N(p^r) \in \mathbb{N}$ such that

$$G \simeq \prod_{p,r} (\mathbb{Z}_{p^r})^{N(p^r)}$$

Theorem 83. If G is finite abelian, then $\forall n, m \in \mathbb{N}$ such that $\#G = nm$, $\exists H \leq G$ with $\#H = n$.

Definition 52. Let X be a set. The *permutation group* of X is the group $S_X \equiv \{\sigma \in X^X \mid \sigma \text{ is a bijection}\}$, under composition of functions. e is the identity function, and σ^{-1} is the inverse function.

Theorem 84. If X and Y have the same cardinality, then $S_X \simeq S_Y$.

Theorem 85. (Cayley) Every group G is isomorphic to a subgroup of S_G.

Theorem 86. Let $\sigma \in S_X$. The relation $x \sim y$ iff $\exists n \in \mathbb{Z} \; \sigma^n(x) = y$ is an equivalence relation on X. The equivalence classes are the *orbits* of σ.

Definition 53. A permutation $\sigma \in S_X$ is a *cycle* iff it has at most one orbit of size greater than 1.

Theorem 87. Every permutation on a finite set is a product of disjoint cycles.

Theorem 88. Let σ be a cycle and O its non-trivial orbit. $\#\langle\sigma\rangle = \#O$.

Theorem 89. If σ, σ' are disjoint cycles, then $\sigma\sigma' = \sigma'\sigma$.

Definition 54. A *transposition* is a cycle of order 2.

Theorem 90. Every non-trivial permutation on a finite set is a product of transpositions.

Definition 55. A permutation is *even/odd* iff it is a product of an even/odd number of transpositions.

Theorem 91. No permutation is both even and odd.

Theorem 92. $A_n \equiv \{\sigma \in S_n \mid \sigma \text{ is even }\} \leq S_n$.

Definition 56. Let G be a group and X a set. A *group action* of G on X is a homomorphism $G \to S_X$.

Definition 57. Under a group action of G on X, the *orbit* of $x \in X$ is $G(x) \equiv \{g(x) \mid g \in G\}$.

Definition 58. Under a group action of G on X,
$G_x \equiv \{g \in G \mid g(x) = x\}$
$G_X \equiv \{g \in G \mid g(x) = x \; \forall x \in X\}$
$X_g \equiv \{x \in X \mid g(x) = x\}$
$X_G \equiv \{x \in X \mid g(x) = x \; \forall g \in G\}$
$X/G \equiv \{G(x) \mid x \in X\}$

Theorem 93. $x \in X_G$ iff $G_x = G$ iff $G(x) = \{x\}$.

Theorem 94. $G_x \leq G$, and $(G : G_x) = \#G(x)$.

Theorem 95. $G_X \trianglelefteq G$

Theorem 96. X/G is a partition of X.

Theorem 97. $G_{g(x)} = gG_xg^{-1}$

Definition 59. A group action is *transitive* iff $\forall x \in X, \; G(x) = X$.

Definition 60. A group action is *faithful* iff $G_X = \{e\}$.

Definition 61. A group action is *free* iff $\forall x \in X, \; G_x = \{e\}$.

Theorem 98. Every free group action is faithful.

Theorem 99. Let X and G be finite. Then

$$\sum_{g \in G} \#X_g = \sum_{x \in X} \#G_x = \#G \cdot \#(X/G)$$

2.2 Ring Theory

Definition 62. A *ring* is an abelian group $(R, +, 0)$ with a function
$\bullet \colon R \times R \to R$ satisfying
1. $(a \cdot b) \cdot c = a \cdot (b \cdot c)$
2. $a \cdot (b + c) = (a \cdot b) + (a \cdot c)$
3. $(a + b) \cdot c = (a \cdot c) + (b \cdot c)$

Facts:

$0 \cdot a = a \cdot 0 = 0$
$(-a) \cdot b = a \cdot (-b) = -(a \cdot b)$

Definition 63. A ring R is *commutative* iff $\forall a, b \in R$, $a \cdot b = b \cdot a$. It
has *unity* or 1 iff $\exists 1 \in R$ such that $\forall a \in R$, $1 \cdot a = a \cdot 1 = a$.

Theorem 100. If $\exists 1' \in R$ such that $\forall a \in R$, $1' \cdot a = a \cdot 1' = a$, then
$1' = 1$.

Theorem 101. If $0 = 1$, then $\#R = 1$.

Theorem 102. In a ring with 1, $\forall n \in \mathbb{Z}$, $na = (n1) \cdot a$.

Theorem 103. Let G be an abelian group. The set of homomor-
phisms $\phi \colon G \to G$, under pointwise addition and composition of functions, is a ring with 1.

Definition 64. A subset $S \subseteq R$ is a *subring* of R, denoted $S \leq R$, iff
it is itself a ring under $+$ and $\bullet$.

Theorem 104. For any two subrings $S, T \leq R$, $S \cap T$ is a subring.

Definition 65. Let R, R' be rings. A function $\phi \colon R \to R'$ is a *homomorphism* iff $\forall a, b \in R$, $\phi(a + b) = \phi(a) + \phi(b)$ and $\phi(a \cdot b) = \phi(a) \cdot \phi(b)$.

Theorem 105. If ϕ is a homomorphism, then:
$\phi(0) = 0$
$\phi(-a) = -\phi(a)$
$S \leq R \Rightarrow \phi(S) \leq R'$
$T \leq R' \Rightarrow \phi^{-1}(T) \leq R$
$\phi(1) = 1$

Definition 66. Let ϕ be a homomorphism. $\ker \phi \equiv \{a \in R \mid \phi(a) = 0\}$.

Theorem 106. $\ker \phi \leq R$

Definition 67. An *isomorphism* is a homomorphism that is bijective. Two rings are *isomorphic*, denoted $R \simeq R'$, iff there exists an isomorphism between them.

Theorem 107. On any set of rings, isomorphism is an equivalence relation.

Theorem 108. If ϕ is an isomorphism, then so is ϕ^{-1}.

Theorem 109. A homomorphism ϕ is injective iff $\ker \phi = \{0\}$.

Definition 68. An *automorphism* is an isomorphism from a ring to itself.

Theorem 110. The set of automorphisms $R \to R$, denoted $\mathrm{Aut}(R)$, is a group under composition of functions.

Theorem 111. For any rings R, S, $R \times S$ is a ring with $(r, s) \cdot (r', s') \equiv (r \cdot r', s \cdot s')$.

Theorem 112. $R \times S \simeq S \times R$

Theorem 113. A subring $N \leq R$ is an *ideal*, denoted $N \trianglelefteq R$, iff $\forall a \in R$, $\forall n \in N$, $a \cdot n \in N$ and $n \cdot a \in N$.

Theorem 114. If ϕ is a homomorphism, then $\ker \phi$ is an ideal.

Theorem 115. Let $N \trianglelefteq R$. R/N is a ring under $(a + N) \cdot (b + N) \equiv (a \cdot b) + N$.

Theorem 116. $R/\ker \phi \simeq \phi(R)$

Theorem 117. If R has 1, then R/N has 1.

Theorem 118. If $S, T \trianglelefteq R$, then $S \cap T \trianglelefteq R$.

Theorem 119. $S, T \trianglelefteq S \times T$, and $(S \times T)/T \simeq S$.

Theorem 120. Let $\phi: R \to R'$ be a homomorphism.
If $N \trianglelefteq R$ then $\phi(N) \trianglelefteq \phi(R)$
If $N' \trianglelefteq \phi(G)$ then $\phi^{-1}(N') \trianglelefteq R$.

Theorem 121. Let $S \leq R$, $N \trianglelefteq R$. Then $S, N \leq S + N \leq R$. If also $S \trianglelefteq R$, then $S + N \trianglelefteq R$.

Theorem 122. Let R be commutative, let $a \in R$. $\{r \in R \mid ar = 0\} \trianglelefteq R$.

Theorem 123. Let R be commutative with 1, let $a \in R$. $\langle a \rangle \equiv \{ra \mid r \in R\} \trianglelefteq R$.

Theorem 124. $\mathbb{Z}$ is a commutative ring with 1, under $(n, m) \cdot (n', m') \equiv (nn' + mm', mn' + nm')$. $(1, 0)$ is unity.

Theorem 125. $n\mathbb{Z} \trianglelefteq \mathbb{Z}$

Definition 69. Let R be a ring with 1. $a \in R$ is a *unit* iff $\exists a^{-1} \in R$ such that $a \cdot a^{-1} = a^{-1} \cdot a = 1$.

Definition 70. $a \in R$, $a \neq 0$ is a *divisor of 0* iff $\exists b \in R$, $b \neq 0$ such that $a \cdot b = 0$.

Theorem 126. $a \in \mathbb{Z}_n$ is a unit, iff $\gcd(a, n) = 1$, iff a is not 0 or a divisor of zero.

Definition 71. An *integral domain* is a commutative ring with no divisors of 0.

Definition 72. A *field* is a commutative ring with 1 such that $\forall a \neq 0$, a is a unit.

Theorem 127. Every field is an integral domain.

Theorem 128. Every finite integral domain is a field.

Theorem 129. $\mathbb{Z}$ is an integral domain.

Theorem 130. Every subring of an integral domain is an integral domain.

Theorem 131. In an integral domain, if $a \cdot b = a \cdot c$, then $b = c$.

Theorem 132. In a ring R, $R^{\times} \equiv \{u \in R \mid u$ is a unit$\}$ is a group under $\bullet$.

Theorem 133. $\forall a \in \mathbb{Z}_n$ such that $\gcd(a, n) = 1$, $a^{\#\mathbb{Z}_n^{\times}} = 1$.

Theorem 134. Let R have 1, $R' \neq \{0\}$, and $\phi \colon R \to R'$ a homomorphism. If $u \in R$ is a unit, then $\phi(u)$ is a unit.

Theorem 135. $a \in R$ is a unit, iff $\langle a \rangle = R$, iff $1 \in \langle a \rangle$.

Theorem 136. For any unit $u \in R$, $i_u \colon R \to R$ such that $i_u(a) = uau^{-1}$ is an automorphism.

Definition 73. Let R be commutative with 1. An ideal $N \trianglelefteq R$ is *prime* iff $\forall a, b \in R$, $a \cdot b \in N \Rightarrow a \in N$ or $b \in N$.

Definition 74. Let R be commutative with 1. An ideal $N \triangleleft R$ is *maximal* iff there is no $M \triangleleft R$ with $N < M < R$.

Theorem 137. Every maximal ideal is prime.

Theorem 138. Let R be commutative with 1. R/N is an integral domain iff N is prime. R/N is a field iff N is maximal.

Definition 75. In a ring R, let $a, b \neq 0$. a *divides* b, denoted $a|b$, iff $\exists c \in R$ such that $b = c \cdot a$.

Theorem 139. In a commutative ring with 1, $a|b$ iff $\langle b \rangle \leq \langle a \rangle$.

Theorem 140. $\langle p \rangle$ is prime iff $p|(a \cdot b) \Rightarrow p|a$ or $p|b$.

Definition 76. $a, b \in R$ are *associated* iff $\exists$ unit $u \in R$ such that $b = u \cdot a$.

Theorem 141. Association is an equivalence relation.

Theorem 142. In an integral domain, a and b are associated iff $\langle a \rangle = \langle b \rangle$.

Definition 77. $p \in R$ is *irreducible* iff $p = a \cdot b \Rightarrow a$ or b is a unit.

Theorem 143. In an integral domain, if $\langle p \rangle$ is prime then p is irreducible.

Definition 78. An integral domain is a *unique factorization domain*, or UFD, iff $\forall a \neq 0$ and not a unit, a is a product of irreducibles, unique up association.

Definition 79. An integral domain D is a *principle ideal domain*, or PID, iff $\forall N \trianglelefteq D$, $\exists a \in D$ such that $N = \langle a \rangle$.

Theorem 144. In a PID, every prime ideal is maximal.

Theorem 145. In a UFD, if p is irreducible then $\langle p \rangle$ is prime.

Theorem 146. In a PID, p is irreducible iff $\langle p \rangle$ is maximal.

Theorem 147. Every PID is a UFD.

Definition 80. A *Boolean* ring is a ring S such that $\forall a \in S$, $a^2 = a$.

Theorem 148. For any set X, $\mathcal{P}(X)$ is a commutative ring under

$$A + B \equiv (A \setminus B) \cup (B \setminus A)$$

$$A \cdot B \equiv A \cap B$$

Theorem 149. In $\mathcal{P}(X)$, $0 = \varnothing$, $A = -A$, and $1 = X$

Theorem 150. A ring is Boolean iff it is isomorphic to a subring of $\mathcal{P}(X)$ for some X.

Definition 81. A *Boolean algebra* is a Boolean ring with 1.

Definition 82. Every finite Boolean algebra is isomorphic to $\mathcal{P}(X)$ for some X.

Theorem 151. $\mathcal{P}(X) \simeq \mathbb{Z}_2^X$

Theorem 152. An ideal of a Boolean algebra is prime iff it is maximal.

Theorem 153. For $S \leq \mathcal{P}(X)$ and $A \subseteq X$, $S_A \equiv \{B \cap A \mid B \in S\} \leq \mathcal{P}(X)$.

Theorem 154. If $A \in S$, then $S_A \trianglelefteq S$, and $S \simeq S_A \times S_{A^c}$.

Definition 83. A *δ-ring* is a Boolean ring S such that $\forall$ countable $\{A_i\} \subseteq S$, $\bigcap_i A_i \in S$.

Definition 84. A *σ-ring* is a Boolean ring S such that $\forall$ countable $\{A_i\} \subseteq S$, $\bigcup_i A_i \in S$.

Theorem 155. Every σ-ring is a δ-ring.

Theorem 156. An intersection of δ-rings is a δ-ring.

Theorem 157. If S is a δ-ring, then so is S_A.

2.3   Field Theory

Definition 85. A *subfield* of a field F, denoted $E \leq F$, is a subring that is a field.

Theorem 158. A field F has no proper nontrivial ideals, only F and $\{0\}$.

Theorem 159. Every field homomorphism is injective.

Theorem 160. $\mathbb{Z}_p$ is a field iff p is prime.

Definition 86. Let D be an integral domain. The *field of fractions* $F(D)$ is the set of equivalence classes in $D^2 \setminus \{(x, 0)\}$ under the equivalence relation $(a, b) \sim (c, d)$ iff $ad = bc$. It is a field under

$$(a, b) + (c, d) = (ad + bc, bd)$$

$$(a, b) \cdot (c, d) = (ac, bd)$$

Theorem 161. D is isomorphic to $\{(x, 1) \in F(D)\}$, a subring of $F(D)$.

Theorem 162. Every field is isomorphic to its own field of fractions.

Theorem 163. Let E be a field and $D \subseteq E$ an integral domain. $F(D)$ is isomorphic to a subfield of E.

Definition 87. The *rational numbers* $\mathbb{Q} \equiv F(\mathbb{Z})$.

Theorem 164. $\mathbb{Q}$ is countable.

Theorem 165. $\mathbb{Q}$ has a natural order, such that $(a, b) < (c, d)$ with $b, d > 0$ iff $ad < bc$.

Theorem 166. $<$ on $\mathbb{Q}$ is dense.

Definition 88. The *characteristic* of a field F, denoted char$F$, is the order of the cyclic subgroup of $+$ generated by 1.

Theorem 167. char$F$ is prime for all fields.

Theorem 168. If char$F = p$ then $\phi: \mathbb{Z}_p \to F$ such that $\phi(n) = n1$ is a homomorphism.

Theorem 169. If char$F$ is infinite, then $\phi: \mathbb{Q} \to F$ such that $\phi(\frac{m}{n}) = (m1)(n1)^{-1}$ is a homomorphism.

Theorem 170. For any field F, $\phi: \mathbb{Z} \to F$ such that $\phi(n) = n1$ is a homomorphism, $\ker \phi = (\text{char}F)\mathbb{Z}$, and $F \geq \phi(\mathbb{Z}) \simeq \mathbb{Z}_{\text{char}F}$.

Theorem 171. char$F = p$ iff $\mathbb{Z}_p \leq F$. char$F$ is infinite iff $\mathbb{Q} \leq F$.

Theorem 172. $\mathbb{Z}_p$ and $\mathbb{Q}$ have no proper subfields.

Definition 89. A field $E \geq F$ is a *finite extension* of F iff it has group structure F^n for some $n \in \mathbb{N}$. Its *dimension* over F is $\dim_F E = n$.

Theorem 173. $\forall p$ prime, $r \in \mathbb{N}$, there is a field F such that $\#F = p^r$.

Theorem 174. If F is finite, then $\exists p$ prime, $r \in \mathbb{N}$ such that $\#F = p^r$.

Theorem 175. If F is finite and $\#E = \#F$, then $E \simeq F$. Let $\mathbb{F}_q$ denote the field of order q.

Theorem 176. $\forall p, r$, $\mathbb{F}_{p^r}$ is a finite extension of $\mathbb{Z}_p$, and $\dim_{\mathbb{Z}_p} \mathbb{F}_{p^r} = r$.

Theorem 177. $\mathrm{char}\mathbb{F}_{p^r} = p$

Theorem 178. $\forall r, n$, $\mathbb{F}_{p^r} \leq \mathbb{F}_{p^{nr}}$

Theorem 179. If F is finite then $F^\times$ is cyclic.

Theorem 180. $\forall a \in \mathbb{F}_q$, $a^q = a$.

Definition 90. A *polynomial* over a field F is a function $p \colon F \to F$ such that for some finite $\{a_n\} \subseteq F$, $p(x) = \sum_n a_n x^n$. The set of polynomials over a field, denoted $P(F)$, is a ring under pointwise addition and multiplication.

Theorem 181. If F is finite, $P(F) = F^F$.

Theorem 182. Let $p \in P(F)$ and $\alpha \in F$. $p(\alpha) = 0$ iff $(x - \alpha)|p$, and $\langle x - \alpha \rangle = \{p \in P(F) \mid p(\alpha) = 0\}$.

Definition 91. Let $E \geq F$. $\alpha \in E$ is *algebraic* over F iff there is a finite, non-zero $\{b_n\} \subseteq F$ such that $\sum_n b_n \alpha^n = 0$.

Definition 92. $E \geq F$ is *algebraic* over F iff every $\alpha \in E$ is algebraic over F.

Theorem 183. Every finite extension of F is algebraic over F.

Theorem 184. Let $E \geq F$, let $\bar{F}_E \equiv \{\alpha \in E \mid \alpha$ algebraic over $F\}$. Then $E \geq \bar{F}_E \geq F$.

Theorem 185. E is algebraic over F iff $\bar{F}_E = E$.

Theorem 186. $\bar{F}_{\bar{F}_E} = \bar{F}_E$

Theorem 187. Let $K \geq E \geq F$. K is algebraic over F iff K is algebraic over E and E is algebraic over F.

Theorem 188. $\overline{(\bar{F}_E)}_E = \bar{F}_E$

Definition 93. A field F is *algebraically closed* iff $\forall p \in P(F)$, $\exists \alpha \in F$ such that $p(\alpha) = 0$.

Theorem 189. F is algebraically closed iff $\forall E$ algebraic over F, $E = F$.

Theorem 190. If $E \geq F$ and F is algebraically closed, then $\bar{F}_E = F$.

Theorem 191. If $E \geq F$ and E is algebraically closed, then $\bar{F}_E$ is algebraically closed.

Theorem 192. If E and F are both algebraically closed and characteristic p, then $E \simeq F$.

Theorem 193. All algebraically closed fields are infinite.

Theorem 194. If E and E' are both algebraic over F and both algebraically closed, then $E \simeq E'$.

Definition 94. An *algebraic closure* of F, denoted $\bar{F}$, is a field that is algebraic over F and algebraically closed.

Theorem 195. Every field has as algebraic closure.

Theorem 196. Let $E \geq F$. Consider the group action of $G = \text{Aut}(E)$. Then $\forall H \leq G$, $E_H \leq E$, and in particular, $F \leq E_{G_F} \leq E$.

Theorem 197. Let $G = \text{Aut}(F)$. $\forall H \leq G$, $F_G \leq F_H$, and $G = G_{F_G}$.

Theorem 198. Let char$F = p$, let $G = \text{Aut}(F)$. $\forall H \leq G$, $\mathbb{Z}_p \leq F_H$, and $G = G_{\mathbb{Z}_p}$.

Theorem 199. Let char$F$ be infinite, let $G = \text{Aut}(F)$. $\forall H \leq G$, $\mathbb{Q} \leq F_H$, and $G = G_{\mathbb{Q}}$.

Theorem 200. Let $E \geq F$, let $G = \text{Aut}(E)$. $\#G_F \leq \dim_F E$.

Definition 95. Let $E \geq F$, let $G = \text{Aut}(E)$. E is a *normal* extension of F, denoted $E \trianglerighteq F$, iff $\#G_F = \dim_F E$.

Theorem 201. $F \trianglerighteq F$

Theorem 202. Let $\bar{F} \geq E \trianglerighteq F$. Then $\forall \sigma \in \text{Aut}(\bar{F})_F$, $\sigma(E) = E$.

Theorem 203. (Galois) Let $K \trianglerighteq F$, let $G = \text{Aut}(K)$. There is a bijection $\lambda \colon \{E \mid K \geq E \geq F\} \to \{H \leq G_F\}$ with $\lambda(E) = G_E$ and $\lambda^{-1}(H) = K_H$.

$K_{G_E} = E$ and $G_{K_H} = H$
$\#\lambda(E) = \dim_E K \quad (K \trianglerighteq E)$
$\dim_F E = (\lambda(F) : \lambda(E))$
$E_1 \geq E_2$ iff $\lambda(E_1) \leq \lambda(E_2)$
$E_1 \trianglerighteq E_2$ iff $\lambda(E_1) \trianglelefteq \lambda(E_2)$
If $E_1 \trianglerighteq E_2$ then $\text{Aut}(E_1)_{E_2} \simeq \lambda(E_2)/\lambda(E_1)$

Theorem 204. If $F \trianglerighteq \mathbb{Z}_p$ then $F_{\text{Aut}(F)} = \mathbb{Z}_p$.

Theorem 205. If $E \geq F \geq \mathbb{Z}_p$ such that E and F are normal over $\mathbb{Z}_p$, then $\text{Aut}(F) \simeq \text{Aut}(E)/\text{Aut}(E)_F$

Theorem 206. If $F \trianglerighteq \mathbb{Q}$ then $F_{\text{Aut}(F)} = \mathbb{Q}$.

Theorem 207. If $E \geq F \geq \mathbb{Q}$ such that E and F are normal over $\mathbb{Q}$, then $\text{Aut}(F) \simeq \text{Aut}(E)/\text{Aut}(E)_F$

Theorem 208. If $\text{char} F$ is infinite, then $\bar{F} \trianglerighteq F$.

Theorem 209. If F is finite then all finite extensions of F are normal.

Theorem 210. If $\#F = q$, and $\dim_F E = n$, then $\text{Aut}(E)_F$ is cyclic of order n, generated by σ such that $\sigma(a) = a^q$.

Theorem 211. If $\#F = p^r$, then $\mathrm{Aut}(F)$ is cyclic of order r, generated by σ such that $\sigma(a) = a^p$.

Definition 96. A field is *real* iff it has an order $<$ satisfying
1. If $a < b$ then $a + c < b + c$
2. If $0 < a, b$ then $0 < ab$
3. $<$ satisfies least upper bound

Theorem 212. All real fields are isomorphic.

Theorem 213. A real field is infinite, with infinite characteristic, and $<$ is dense.

Definition 97. (Dedekind cuts) Define the subset $\mathbb{R} \subseteq \mathcal{P}(\mathbb{Q})$ such that $X \in \mathbb{R}$ iff
1. $X \neq \varnothing$ and $X \neq \mathbb{Q}$
2. $x \in X \Rightarrow \forall y < x, \ y \in X$
3. $x \in X \Rightarrow \exists y \in X$ such that $y > x$

Define an order on $\mathbb{R}$ by $X < Y$ iff $X \subset Y$.
Let $0 \equiv \{x < 0\}$.
For $X, Y \in \mathbb{R}$, define $X + Y \equiv \{z \in \mathbb{Q} \mid z = x + y \ \exists x \in X \ \exists y \in Y\}$.
For $X, Y \in \mathbb{R}$, $X, Y > 0$, define $XY \equiv \{z \in \mathbb{Q} \mid z < xy \ \exists x \in X \ x > 0, \ \exists y \in Y \ y > 0\}$.
Extend this multiplication to $\mathbb{R}$ by $0X = X0 = 0$, $(-X)Y = X(-Y) = -XY$, and $(-X)(-Y) = XY$.

Theorem 214. $\mathbb{R}$ is a real field.

Theorem 215. $\mathbb{R}$ is uncountable.

Theorem 216. $\mathrm{Aut}(\mathbb{R}) = \{e\}$

Theorem 217. $\mathbb{Q} \simeq \{\{x < q\} \mid q \in \mathbb{Q}\} \leq \mathbb{R}$

Definition 98. The *complex numbers* are the field $\mathbb{C} \equiv \mathbb{R}^2$ with $(a, b) + (c, d) \equiv (a + c, b + d)$ and $(a, b) \cdot (c, d) \equiv (ac - bd, ad + bc)$. $i \equiv (0, 1)$.

Theorem 218. $\mathbb{C} = \bar{\mathbb{R}}$

Theorem 219. $\mathbb{R} \simeq \{(x,0) \mid x \in \mathbb{R}\} \leq \mathbb{C}$

Definition 99. *Complex conjugation* is the function $\mathbb{C} \to \mathbb{C}$ given by $\overline{(a,b)} = (a,-b)$.

Theorem 220. $\forall z \in \mathbb{C}, \bar{\bar{z}} = z$.

Theorem 221. Complex conjugation is an automorphism.

2.4   Linear Algebra

Definition 100. A *vector space* over a field F is an abelian group $(V,+)$ with a scalar multiplication $F \times V \to V$ satisfying $\forall \alpha, \beta \in F, \forall v, w \in V$,
1. $\alpha(v + w) = \alpha v + \alpha w$
2. $(\alpha + \beta)v = \alpha v + \beta v$
3. $\alpha(\beta v) = (\alpha\beta)v$
4. $1v = v$

A *vector* is an element of a vector space.

Facts:

$0v = 0$
$(-\alpha)v = -(\alpha v)$

Theorem 222. Any field is a vector space over itself.

Definition 101. A *linear combination* of vectors $\{v_i\}$ is a vector of the form

$$\alpha_1 v_1 + \dots + \alpha_n v_n$$

where $\alpha_i \in F$.

Definition 102. A set of vectors $S \subseteq V$ is *linearly independent* iff none of them can be written as a finite linear combination of the others, equivalently, for any $\{v_i\} \subseteq S$,

$$\alpha_1 v_1 + ... + \alpha_n v_n = 0 \;\Rightarrow\; \forall i, \; \alpha_i = 0$$

A set is *linearly dependent* iff it is not linearly independent.

Definition 103. A *subspace* $W \leq V$ is a subset which is itself a vector space.

Definition 104. The *span* of a set of vectors $S \subseteq V$, written span$S$, is the smallest subspace of V containing S, equivalently, it is the space of all finite linear combinations of elements S.

Definition 105. $S \subseteq V$ is a (Hamel/algebraic) *basis* iff it is linearly independent and span$S = V$.

Theorem 223. If S is a basis, every element of V is a unique finite linear combination of S.

Theorem 224. Every vector space has a basis.

Theorem 225. If S is linearly independent then $\exists S' \supseteq S$ such that S' is a basis. If span$S = V$ then $\exists S' \subseteq S$ such that S' is a basis.

Theorem 226. span span$S = $ span$S$

Theorem 227. If $S \subseteq S'$ then span$S \subseteq$ spanS'.

Theorem 228. Let $S \subseteq S'$. If S' is linearly independent then so is S. If S is linearly dependent then so is S'.

Theorem 229. All bases of V have the same cardinality.

Definition 106. A vector space is *finite dimensional* if it has a finite basis. The *dimension* of V, written $\dim V$, is the cardinality of its basis.

Theorem 230. If S is linearly independent and $\#S = \dim V$, then S is a basis. If $\mathrm{span} S = V$ and $\#S = \dim V$, then S is a basis.

Theorem 231. If $W \leq V$ then $\dim W \leq \dim V$.

Definition 107. Let V, W be vector spaces over F. A function $T \colon V \to W$ is a *linear map* iff $\forall \alpha, \beta \in F, \ \forall v, w \in V,$

$$T(\alpha v + \beta w) = \alpha T(v) + \beta T(w)$$

The space of all linear maps $V \to W$ is denoted $\mathcal{L}(V, W)$.

Theorem 232. $\mathcal{L}(V, W)$ is a vector space.

Definition 108. An *algebra* A is a vector space and a ring, satisfying $\forall \alpha \in F, \ \forall T, S \in A,$

$$\alpha(TS) = (\alpha T)S = T(\alpha S)$$

Theorem 233. $\mathcal{L}(V) \equiv \mathcal{L}(V, V)$ is an algebra with 1 under composition of functions.

Definition 109. A function $f \colon V \to W$ is *affine* iff it has the form $f(v) = L(v) + c$, where $L \in \mathcal{L}(V, W)$ and $c \in W$.

Definition 110. A function is an *isomorphism* iff it is linear and a bijection. Two vector spaces are *isomorphic*, denoted $V \simeq V'$, iff there exists an isomorphism between them.

Theorem 234. On any set of vector spaces, isomorphism is an equivalence relation.

Definition 111. The group of isomorphisms in $\mathcal{L}(V)$ is called the *general linear* group, and denoted $GL(V)$.

Theorem 235. Let $T \in \mathcal{L}(V, W)$. Then $T(V) \leq W$, $\ker T \leq V$, and $\dim V = \dim(\ker(T)) + \dim T(V)$.

Theorem 236. Let $T \in \mathcal{L}(V, W)$. If S is a basis for V, then $\mathrm{span} T(S) = T(V)$, and $\dim T(V) \leq \dim V$.

Theorem 237. Let $T \in \mathcal{L}(V, W)$ be an isomorphism. Then S is a basis for V iff $T(S)$ is a basis for W, and $\dim V = \dim W$.

Theorem 238. Let V be finite dimensional. Let $T \in \mathcal{L}(V)$. T is injective iff it is also surjective.

Theorem 239. V and W over F are isomorphic iff $\dim V = \dim W$.

Theorem 240. $\dim V = n$ iff V is isomorphic to F^n with the natural addition and scalar multiplication.

Theorem 241. V is naturally isomorphic to $\mathcal{L}(F, V)$ under the action $v(\alpha) = \alpha v$.

Theorem 242. An arbitrary product of vector spaces over F is a vector space over F, under pointwise addition and scalar multiplication.

Definition 112. The *direct sum* of vector spaces A and B over F is $A \oplus B \equiv A \times B$ with the natural addition and scalar multiplication.

Definition 113. Let $\{V_i\}_{i \in I}$ be an indexed family of vector spaces over F. Define $\bigoplus V_i$ to be the subspace of $\prod V_i$ such that $v \in \bigoplus V_i$ iff $v_i = 0$ for all but finitely many $i \in I$.

Theorem 243. For $i \in I$, define $e_i \colon I \to F$ as

$$e_i(j) \equiv \begin{cases} 1 & i = j \\ 0 & i \neq j \end{cases}$$

Then $\{e_i\}$ is a basis for $\bigoplus_{i \in I} F$.

Definition 114. If $A, B \leq V$, define $A + B \equiv \operatorname{span}(A \cup B)$.

Theorem 244. Let $A, B \leq V$. $A \cap B$ is a vector space.

Theorem 245. Let $W \leq V$. V/W is a vector space.

Facts:

$A \cap B \leq A, B \leq A + B \leq V$
$A, B \leq A \oplus B$
$A + B = A \oplus B$ iff $A \cap B = \{0\}$
$\dim(A \oplus B) = \dim A + \dim B$
$\dim(V/W) = \dim V - \dim W$
$\dim(A + B) = \dim A + \dim B - \dim(A \cap B)$

Theorem 246. Let $E \geq F$ be fields. Let V be a vector space over E. Then V is also a vector space over F, and

$$\dim_F V = \dim_E V \cdot \dim_F E$$

Definition 115. Let $T \in \mathcal{L}(V)$. $\lambda \in F$ is an *eigenvalue* of T iff $\exists v \neq 0$ such that $T(v) = \lambda v$.

Theorem 247. $V_T(\lambda) \equiv \{v \in V \mid T(v) = \lambda v\} \leq V$.

Theorem 248. If $\lambda_1 \neq \lambda_2$ are eigenvalues of T, then $V_T(\lambda_1) \cap V_T(\lambda_2) = \{0\}$.

Definition 116. Let V be a vector space over F. Its *dual space* is $V^* \equiv \mathcal{L}(V, F)$.

Theorem 249. If V is finite dimensional, then it is isomorphic to V^*.

Theorem 250. Define the function $i\colon V \to V^{**}$ such that $\forall \omega \in V^*$, $i(v)(\omega) = \omega(v)$. i is an injective linear map. If V is finite dimensional, then i is an isomorphism.

Theorem 251. Let V be finite dimensional. If $\{e_i\}$ is a basis for V, then there exists a unique basis $\{e^i\}$ of V^* satisfying $e^i(e_j) = \delta^i{}_j$, where

$$\delta^i{}_j = \begin{cases} 1 & i = j \\ 0 & i \neq j \end{cases}$$

Theorem 252. $(A \oplus B)^* = A^* \oplus B^*$

Definition 117. Let V, W be vector spaces over F, and let F have complex conjugation. A function $T\colon V \to W$ is *antilinear* iff $\forall \alpha, \beta \in F$, $\forall v, w \in V$,
$$T(\alpha v + \beta w) = \bar{\alpha} T(v) + \bar{\beta} T(w)$$

Definition 118. Let V be a vector space over F. If F has complex conjugation, then the *conjugate space* $\bar{V}$ is the space of antilinear maps $v\colon F \to V$.

Definition 119. Let $v \in V$. The *complex conjugate* of v is $\bar{v} \in \bar{V}$ such that $\forall \alpha$, $\bar{v}(\alpha) = \bar{\alpha} v$.

Theorem 253. Complex conjugation is an antilinear isomorphism.

Definition 120. Let $\dim V = n$. The *trace* is the unique linear map $\mathrm{Tr} \in \mathcal{L}(V)^*$ satisfying
1. $\mathrm{Tr}(AB) = \mathrm{Tr}(BA)$
2. $\mathrm{Tr}(\mathbf{1}) = n$

Theorem 254. For any basis $\{e_i\}$, $\mathrm{Tr}(A) = \sum e^i(A(e_i))$.

Theorem 255. If V and W are finite dimensional, then $\mathcal{L}(V, W)^* = \mathcal{L}(W, V)$ For $A \in \mathcal{L}(V, W)$, and $B \in \mathcal{L}(W, V)$, the pairing is given by $\mathrm{Tr}(A \circ B) = \mathrm{Tr}(B \circ A)$.

Definition 121. Let $L \in \mathcal{L}(V, W)$. The *transpose* of L is the map $L^T \in \mathcal{L}(W^*, V^*)$ such that $\forall v \in V$, $\forall \omega \in W^*$,
$$L^T(\omega)(v) = \omega(L(v))$$

Theorem 256. Transpose is an injective linear map. If V and W are finite dimensional, then transpose is an isomorphism.

Facts:

$(A \circ B)^T = B^T \circ A^T$
$A^{TT} = A$
$\mathbf{1}^T = \mathbf{1}$
$(A^{-1})^T = (A^T)^{-1}$
$\mathrm{Tr}(A^T) = \mathrm{Tr} A$

Theorem 257. Let $L \in \mathcal{L}(V, W)$.

$$\ker L^T = (W/L(V))^*$$

$$L^T(W^*) = (V/\ker L)^*$$

$$(\ker L)^* = V^*/L^T(W^*)$$

$$L(V)^* = W^*/\ker L^T$$

Definition 122. A vector function $T \colon \prod V_i \to W$ is *multilinear* iff it is linear on each of its arguments separately.

Definition 123. Let V, W be finite dimensional vector spaces over F. Their *tensor product* $V \otimes W$ is the space of bilinear maps $T \colon V^* \times W^* \to F$.

Facts:

$\dim(V \otimes W) = \dim V \cdot \dim W$
$(V \otimes W)^* = W^* \otimes V^*$
$A \otimes (B \oplus C) = (A \otimes B) \oplus (A \otimes C)$

Definition 124. Let V be finite dimensional over F. A *tensor* of type (l, k, n, m) over V is a multilinear map

$$T \colon \underbrace{V^* \times \cdots \times V^*}_{l} \times \underbrace{V \times \cdots \times V}_{k} \times \underbrace{\bar{V}^* \times \cdots \times \bar{V}^*}_{n} \times \underbrace{\bar{V} \times \cdots \times \bar{V}}_{m} \to F$$

It is written with l upper indices, k lower indices, n upper primed indices, and m lower primed indices, such as $T^{aa'}{}_{bb'}$. The set of such tensors is denoted $\mathcal{T}(l, k, n, m)$. The n and m designations can be omitted if not needed.

Theorem 258. $\mathcal{T}(l, k, n, m)$ is a vector space of dimension $(\dim V)^{l+k+n+m}$.

Theorem 259. $\mathcal{T}(l, k, n, m)^* = \mathcal{T}(k, l, m, n)$. $\overline{\mathcal{T}(l, k, n, m)} = \mathcal{T}(n, m, l, k)$.

Definition 125. The *complex conjugate* of a tensor T of type (l, k, n, m) is the tensor $\bar{T}$ of type (n, m, l, k) given by

$$\bar{T}(\omega_i, v_j, \psi_p, w_q) = \overline{T(\bar{\psi}_p, \bar{w}_q, \bar{\omega}_i, \bar{v}_j)}$$

Theorem 260. Complex conjugation of tensors is an antilinear isomorphism.

Theorem 261. $\mathcal{T}(1, 1)$ is isomorphic to $\mathcal{L}(V)$, under the natural isomorphism by which a linear map L acts as a tensor by $L(\omega, v) = \omega(L(v))$.

Definition 126. The *tensor product* is the bilinear map $\otimes\colon \mathcal{T}(l, k, n, m) \times \mathcal{T}(l', k', n', m') \to \mathcal{T}(l + l', k + k', n + n', m + m')$ given by

$$(T \otimes S)(v_i, w_j) = T(v_i) \cdot S(w_j)$$

Definition 127. The *contraction* of the ith upper and jth lower indices is a linear map $\mathcal{T}(l, k) \to \mathcal{T}(l - 1, k - 1)$ given by letting the tensor act on all of its arguments except the ith upper and jth lower, and taking the trace of the resulting tensor of type $(1, 1)$. It is denoted by writing the ith upper and jth lower indices with the same letter, such as $T^a{}_a$. A similar definition applies to contracting an upper primed index with a lower primed index.

Theorem 262. Let $\{e_i\}$ be a basis for V.

$$\{\underbrace{e_i \otimes \cdots \otimes e_j}_{l} \otimes \underbrace{e^n \otimes \cdots \otimes e^m}_{k}\}$$

is a basis for $\mathcal{T}(l, k)$.

Theorem 263. Let $\{e_i\}$ be a basis for V.

$$\sum e_i \otimes e^i = \mathbf{1}$$

Theorem 264. $GL(V)$ has a natural action on $\mathcal{T}(l, k)$ which preserves contraction, given by

$$L(T) = L^{a_1}{}_{b_1} \cdots L^{a_l}{}_{b_l} \; T^{b_1 \ldots b_l}{}_{c_1 \ldots c_k} \; L^{-1}{}^{c_1}{}_{d_1} \cdots L^{-1}{}^{c_k}{}_{d_k}$$

Definition 128. A tensor T of type $(l, 0)$ is *symmetric* iff $T(\omega_i) = T(\sigma(\omega_i))$ for all permutations σ. The space of such tensors is denoted $\mathrm{Sym}\mathcal{T}(l, 0)$. T is *antisymmetric* iff $T(\omega_i) = T(\sigma(\omega_i))$ for all even permutations σ, and $T(\omega_i) = -T(\sigma(\omega_i))$ for all odd permutations σ. The space of such tensors is denoted $\mathrm{Asym}\mathcal{T}(l, 0)$. A similar definition holds for lower indices.

Definition 129. The *symmetric part* of a tensor $T \in \mathcal{T}(l, 0)$ is $\mathrm{Sym}T \in \mathrm{Sym}\mathcal{T}(l, 0)$ given by

$$\mathrm{Sym}T(\omega_i) = \frac{1}{l!} \sum_\sigma T(\sigma(\omega_i))$$

It is also written $T^{(abc\ldots)}$. The *antisymmetric part* of T is $\mathrm{Asym}T \in \mathrm{Asym}\mathcal{T}(l, 0)$ given by

$$\mathrm{Asym}T(\omega_i) = \frac{1}{l!} \sum_\sigma (-1)^\sigma T(\sigma(\omega_i))$$

It is also written $T^{[abc\ldots]}$. A similar definition holds for lower indices.

Definition 130. Let $\dim V = n$. A *volume form* is a tensor $\epsilon_{a_1 \ldots a_n} \in \mathrm{Asym}\mathcal{T}(0, n)$ such that $\epsilon_{a_1 \ldots a_n} \neq 0$.

Theorem 265. $\dim \mathrm{Asym}\mathcal{T}(0, n) = \dim \mathrm{Asym}\mathcal{T}(n, 0) = 1$. Given a volume form $\epsilon_{a_1 \ldots a_n}$, there exists a unique tensor $\epsilon^{a_1 \ldots a_n} \in \mathrm{Asym}\mathcal{T}(n, 0)$ such that

$$\epsilon_{a_1 \ldots a_n} \epsilon^{a_1 \ldots a_n} = n!$$

Theorem 266. Let $\delta^a_{\ b}$ denote the identity linear map.

$$\epsilon_{a_1 \ldots a_n} \epsilon^{b_1 \ldots b_n} = n!\, \delta^{b_1}_{[a_1} \cdots \delta^{b_n}_{a_n]}$$

Definition 131. Let $L \in \mathcal{L}(V)$. The *determinant* of L is the unique number $\det L \in F$ such that

$$\epsilon_{a_1 \ldots a_n} L^{a_1}{}_{b_1} \cdots L^{a_n}{}_{b_n} = \det L\ \epsilon_{b_1 \ldots b_n}$$

Theorem 267. The determinant is independent of volume form.

Facts:

$\det(AB) = \det A \det B$
$\det \mathbf{1} = 1$
$\det 0 = 0$
$\det(\alpha L) = \alpha^n \det L$
L is invertible iff $\det L \neq 0$
$\det(A^T) = \det A$

Definition 132. The *special linear* group is the subgroup $SL(V) \leq GL(V)$ consisting of the linear maps L with $\det L = 1$.

Definition 133. If V is a vector space over $\mathbb{R}$, an *inner product* on V is a tensor g_{ab} of type $(0,2)$, denoted $\forall v, w \in V$, $\langle v, w \rangle \equiv v^a g_{ab} w^b$, satisfying

1. (symmetric) $\langle v, w \rangle = \langle w, v \rangle$
2. (positive definite) $\langle v, v \rangle \geq 0$, and $\langle v, v \rangle = 0$ iff $v = 0$.

If V is a vector space over $\mathbb{C}$, an *inner product* on V is a tensor $g_{a'b}$ of type $(0,1,0,1)$, denoted $\forall v, w \in V$, $\langle v, w \rangle \equiv \bar{v}^{a'} g_{a'b} w^b$, satisfying

1. (Hermitian) $\langle v, w \rangle = \overline{\langle w, v \rangle}$
2. (positive definite) $\langle v, v \rangle \geq 0$, and $\langle v, v \rangle = 0$ iff $v = 0$.

Theorem 268. Every finite-dimensional vector space over $\mathbb{R}$ or $\mathbb{C}$ has an inner product.

Definition 134. Let V be a space with an inner product. The *norm* of $v \in V$ is
$$\|v\| \equiv \sqrt{\langle v, v \rangle}$$

Theorem 269. (Cauchy-Schwartz) $|\langle v, w \rangle| \leq \|v\| \|w\|$

Definition 135. For $v, w \in V$ over $\mathbb{R}$, the *angle* between them is the unique $\theta \in [0, \pi]$ such that

$$\cos\theta = \frac{\langle v, w \rangle}{\|v\|\|w\|}$$

Definition 136. $v, w \in V$ are *orthogonal* iff $\langle v, w \rangle = 0$. A set of vectors $\{v_i\}$ is *orthonormal* iff $\langle v_i, v_j \rangle = \delta_{ij}$.

Theorem 270. If a set is orthonormal then it is linearly independent.

Definition 137. Let $v \in V$. The *projection* onto v is the linear map P_v given by

$$P_v(w) = \frac{\langle v, w \rangle v}{\|v\|^2}$$

Definition 138. For $v \in V$, the *adjoint* of v is the unique dual vector $v^\dagger \in V^*$ such that $\forall w$, $v^\dagger(w) = \langle v, w \rangle$. Equivalently, $v^\dagger_a = v^b g_{ba}$ or $v^\dagger_a = \bar{v}^{b'} g_{b'a}$.

Theorem 271. If V is finite dimensional, then adjoint is an antilinear isomorphism between V and V^*. Equivalently, there exists a unique tensor g^{ab} such that $g^{ab}g_{bc} = \delta^a{}_c$.

Theorem 272. If a basis $\{e_i\}$ is orthonormal, then $e^i = e^\dagger_i$, and $\forall v$,

$$v = \sum \langle e_i, v \rangle e_i$$

Theorem 273. If V has an inner product, then $\mathcal{T}(l, k)$ acquires a natural inner product given by

$$\langle T, S \rangle \equiv T^{a_1 \ldots a_l}{}_{b_1 \ldots b_k} \; g_{a_1 c_1} \cdots g_{a_l c_l} g^{b_1 d_1} \cdots g^{b_k d_k} \; S^{c_1 \ldots c_l}{}_{d_1 \ldots d_k}$$

Definition 139. Let V be finite dimensional and let $L \in \mathcal{L}(V)$. The *adjoint* of L is the unique linear map $L^\dagger \in \mathcal{L}(V)$ such that $\forall v, w \in V$,

$$\langle v, L(w) \rangle = \langle L^\dagger(v), w \rangle$$

Equivalently, $L^\dagger{}^a{}_b = g^{ac}L^d{}_c g_{db}$, or $L^\dagger{}^a{}_b = g^{ac'}\bar{L}^{d'}{}_{c'} g_{d'b}$.

Theorem 274. Adjoint is an antilinear isomorphism of $\mathcal{L}(V)$ with itself.

Facts:

$(AB)^\dagger = B^\dagger A^\dagger$
$A^{\dagger\dagger} = A$
$\mathbf{1}^\dagger = \mathbf{1}$
$(A^{-1})^\dagger = (A^\dagger)^{-1}$
$\text{Tr}(A^\dagger) = \overline{\text{Tr} A}$
$\det(A^\dagger) = \overline{\det A}$

Theorem 275. $L^T(v^\dagger) = (L^\dagger(v))^\dagger$

Definition 140. $L \in \mathcal{L}(V)$ is *Hermitian* iff $L = L^\dagger$.

Theorem 276. If L is Hermitian then the eigenvalues of L are real, and if $\lambda_1 \neq \lambda_2$ then $V_L(\lambda_1)$ is orthogonal to $V_L(\lambda_2)$.

Definition 141. Let V be real (complex). $L \in \mathcal{L}(V)$ is *orthogonal (unitary)* iff $\forall v, w \in V$,

$$\langle L(v), L(w) \rangle = \langle v, w \rangle$$

Equivalently, $L^\dagger L = \mathbf{1}$.

Theorem 277. If L is orthogonal (unitary), then $|\det L| = 1$.

Theorem 278. The orthogonal (unitary) maps form a subgroup of $GL(V)$, denoted $O(V, g)$ ($U(V, g)$).

Theorem 279. Let $\dim V = n$, $\dim W = m$. Let $A \in \mathcal{L}(V)$, $B \in \mathcal{L}(W)$. Then $A \otimes B \in \mathcal{L}(V \otimes W)$. λ is an eigenvalue of $A \otimes B$ iff $\lambda = \lambda_1 \cdot \lambda_2$, where λ_1, λ_2 are eigenvalues of A, B respectively.

Facts:

$\text{Tr}(A \otimes B) = \text{Tr} A \cdot \text{Tr} B$
$\det(A \otimes B) = (\det A)^m (\det B)^n$
$(A \otimes B)^{-1} = A^{-1} \otimes B^{-1}$
$(A \otimes B)^T = B^T \otimes A^T$
$(A \otimes B)^\dagger = A^\dagger \otimes B^\dagger$

Chapter 3

Topology

Definition 142. Let X be a set. A *topology* on X is a set $\tau \subseteq \mathcal{P}(X)$ satisfying
1. $\varnothing \in \tau$ and $X \in \tau$
2. $U, V \in \tau \Rightarrow U \cap V \in \tau$
3. $S \subseteq \tau \Rightarrow \bigcup S \in \tau$

A *topological space* is a set with a topology.

Theorem 280. If τ_1 and τ_2 are topologies, then $\tau_1 \cap \tau_2$ is a topology.

Definition 143. $U \subseteq X$ is *open* iff $U \in \tau$. $F \subseteq X$ is *closed* iff $F^c \in \tau$.

Definition 144. The *discrete* topology on X is $\tau = \mathcal{P}(X)$

Definition 145. The *trivial* topology on X is $\tau = \{\varnothing, X\}$.

Definition 146. A topology *base* is $B \subseteq \mathcal{P}(X)$ satisfying
1. $\forall x \in X$, $\exists b \in B$ with $x \in b$
2. If $b_1, b_2 \in B$ and $x \in b_1 \cap b_2$ then $\exists b_3 \in B$ with $x \in b_3$ and $b_3 \subseteq b_1 \cap b_2$

Theorem 281. If B is a base on X, then $\{\bigcup S \mid S \subseteq B\}$ is a topology on X.

Theorem 282. B is a base for topology τ iff $\forall U \in \tau$, $\forall x \in U$, $\exists b \in B$ with $x \in b \subseteq U$.

Theorem 283. Let B be a base of τ. $U \in \tau$ iff $\forall x \in U$, $\exists b \in B$ with $x \in b \subseteq U$.

Definition 147. Let X be a set with an order $<$. The *order topology* on X is that generated from the base $B_<$ such that $b \in B_<$ iff either:
1. $\exists x \in X$ $b = \{z \mid z < x\}$, or
2. $\exists y \in X$ $b = \{z \mid z > y\}$, or
3. $\exists x, y \in X$ $b = \{z \mid x < z < y\}$

Definition 148. Let X have topology τ, let $A \subseteq X$. The *subspace topology* on A is $\tau_A \equiv \{A \cap U \mid U \in \tau\}$.

Definition 149. Let $\{X_\alpha\}$ be an indexed family of topological spaces. The *product topology* on $\prod_\alpha X_\alpha$ is that generated from the base $B \equiv \{\prod U_\alpha \mid U_\alpha \in \tau_\alpha$, and $U_\alpha = X_\alpha$ for all but finitely many $\alpha\}$.

Theorem 284. Let $A \subseteq X$, $B \subseteq Y$, then on $A \times B$, the product subspace topology equals the subspace product topology.

Definition 150. Let X be a topological space, and let P be a partition of X. The *quotient topology* on P is $\tau_P \equiv \{U \subseteq P \mid \bigcup U$ is open$\}$.

Definition 151. For any set X, the *power topology* on $\mathcal{P}(X)$ is the product topology on $2^X = \mathcal{P}(X)$, where 2 is discrete.

Definition 152. Let X be a topological space, and $A \subseteq X$. The *interior* of A is $\mathring{A} \equiv \bigcup \{U \mid U$ is open$, U \subseteq A\}$. The *closure* of A is $\bar{A} \equiv \bigcap \{F \mid F$ is closed$, F \supseteq A\}$.

Theorem 285. If U is open and $U \subseteq A$, then $U \subseteq \mathring{A}$. If F is closed and $F \supseteq A$, then $F \supseteq \bar{A}$.

Definition 153. Let X be a topological space and let $x \in X$. A *neighborhood* of x is a $U \in \tau$ with $x \in U$.

Theorem 286. $x \in \mathring{A}$ iff x has a neighborhood contained in A.

Theorem 287. $x \in \bar{A}$ iff every neighborhood of x intersects A.

Theorem 288. A is open iff $A = \mathring{A}$. A is closed iff $A = \bar{A}$.

Definition 154. The *boundary* of A is $\partial A \equiv \bar{A} \cap (\mathring{A})^c$.

Facts:

$\mathring{A} \subseteq A$

$\mathring{\mathring{A}} = \mathring{A}$

$\mathring{A}$ is open

$\bar{A} \supseteq A$

$\bar{\bar{A}} = \bar{A}$

$\bar{A}$ is closed

$A \subseteq B \Rightarrow \mathring{A} \subseteq \mathring{B}$

$A \subseteq B \Rightarrow \bar{A} \subseteq \bar{B}$

$\mathring{A} \cup \mathring{B} \subseteq (A \cup B)^\circ$

$\mathring{A} \cap \mathring{B} = (A \cap B)^\circ$

$\bar{A} \cup \bar{B} = \overline{A \cup B}$

$\bar{A} \cap \bar{B} \supseteq \overline{A \cap B}$

$\bar{A} \times \bar{B} = \overline{A \times B}$

$\mathring{A} \times \mathring{B} = (A \times B)^\circ$

$\bar{A}^c = (A^c)^\circ$

$\overline{A^c} = (\mathring{A})^c$

$\mathring{\bar{\mathring{A}}} = \mathring{A}$

$\overline{\mathring{\bar{A}}} = \bar{A}$

$\partial(A \cup B) \subseteq \partial A \cup \partial B$

$\partial \bar{A} \subseteq \partial A$

$\partial \mathring{A} \subseteq \partial A$

$\bar{A} = \mathring{A} \cup \partial A$

$\mathring{A} = \bar{A} \setminus \partial A$

A is closed iff $\partial A \subseteq A$

A is open iff $\partial A \cap A = \varnothing$

$\partial A = \partial(A^c)$

∂A is closed

Definition 155. A point $x \in X$ is *isolated* iff $\{x\}$ is open.

Definition 156. A space is *dense* iff it has no isolated points.

Definition 157. $A \subseteq X$ is *dense* in X iff $\bar{A} = X$.

Definition 158. x is a *limit point* of A iff every neighborhood U of x satisfies $(U \setminus \{x\}) \cap A \neq \varnothing$.

Definition 159. A *sequence* in X is an indexed family $\{x_n\} \in X^{\mathbb{N}}$. A sequence $\{x_n\}$ *converges* to $x \in X$, denoted $x = \lim x_n$, iff for every neighborhood U of x, $\exists N \in \mathbb{N}$ such that $\forall n \geq N,\ x_n \in U$.

Definition 160. A *sub-permutation* of a sequence $\{x_n\}$ is a sequence $\{x_{\sigma(n)}\}$ where $\sigma \colon \mathbb{N} \to \mathbb{N}$ is injective.

Theorem 289. If $\{x_n\}$ converges to x, then every sub-permutation of $\{x_n\}$ converges to x.

Theorem 290. For $A \subseteq X$, if there is a sequence $\{a_n\} \subseteq A$ converging to $x \in X$, then $x \in \bar{A}$.

Definition 161. A topological space X is *Hausdorff* iff $\forall x, y \in X$ with $x \neq y$, $\exists U_x, U_y \in \tau$ such that $x \in U_x$, $y \in U_y$, and $U_x \cap U_y = \varnothing$.

Theorem 291. Let X be Hausdorff. If a sequence $\{x_n\}$ converges to x and to y, then $x = y$.

Theorem 292. A sequence $\{\{x_\alpha\}_n\} \subseteq \prod X_\alpha$ converges to $\{y_\alpha\}$ iff $\forall \alpha$, the sequence $\{x_{\alpha n}\} \subseteq X_\alpha$ converges to y_α.

Definition 162. A function $f \colon X \to Y$ between topological spaces is *continuous* iff $\forall U \subseteq Y$ open, $f^{-1}(U)$ is open.

Theorem 293. The following are equivalent:
1. $f \colon X \to Y$ is continuous.
2. For a base B on Y, $\forall b \in B,\ f^{-1}(b)$ is open.
3. $\forall x \in X$, for every open $U \ni f(x)$, there is an open $V \ni x$ with $f(V) \subseteq U$.
4. $\forall A \subseteq X,\ f(\bar{A}) \subseteq \overline{f(A)}$.
5. $\forall F \subseteq Y$ closed, $f^{-1}(F)$ is closed.

Theorem 294. Any function from the discrete topology is continuous. Any function to the trivial topology is continuous.

Theorem 295. The identity function is continuous.

Theorem 296. A constant function is continuous.

Theorem 297. If $f\colon X \to Y$ and $g\colon Y \to Z$ are continuous, then $g \circ f$ is continuous.

Theorem 298. If f, g are continuous, then $f \times g$ is continuous.

Theorem 299. A function $f\colon X \to \prod Y_\alpha$ is continuous iff $\forall \alpha$, $f_\alpha\colon X \to Y_\alpha$ is continuous.

Theorem 300. If f is continuous and $\{x_n\}$ converges to x, then $\{f(x_n)\}$ converges to $f(x)$.

Theorem 301. (Pasting) Let $X = A \cup B$ with A, B closed. Let $f\colon A \to Y$, $g\colon B \to Y$ be continuous and $\forall x \in A \cap B$, $f(x) = g(x)$. Then $h\colon X \to Y$ such that

$$h(x) = \begin{cases} f(x) & x \in A \\ g(x) & x \in B \end{cases}$$

is continuous.

Theorem 302. Let Y be Hausdorff, let $A \subseteq X$ be dense in X. If $f, g\colon X \to Y$ are continuous and $f(a) = g(a)$ $\forall a \in A$, then $f = g$.

Definition 163. A function f is a *homeomorphism* iff it is bijective and both f and f^{-1} are continuous. Topological spaces are *homeomorphic* iff there exists a homeomorphism between them.

Definition 164. On any set of topological spaces, homeomorphism is an equivalence relation.

Definition 165. A space X is *homogeneous* iff $\forall x, y \in X$, there is a homomorphism $f\colon X \to X$ with $f(x) = y$.

Definition 166. A *separation* of X is a pair $U, V \subseteq X$ open and non-empty, such that $U \cap V = \varnothing$ and $U \cup V = X$.

Definition 167. A space is *connected* iff it has no separation.

Theorem 303. If $f \colon X \to Y$ is continuous and X is connected, then $f(X)$ is connected.

Theorem 304. The discrete topology is not connected. The trivial topology is connected.

Theorem 305. If $A \subseteq X$ is connected, and $X = U \cup V$ is a separation, then $A \subseteq U$ or $A \subseteq V$.

Theorem 306. If $X \supseteq A \cap B \neq \varnothing$, and A and B are connected, then $A \cup B$ is connected.

Theorem 307. If $A \subseteq X$ is connected, and $A \subseteq B \subseteq \bar{A}$, then B is connected.

Theorem 308. Every product of connected spaces is connected.

Theorem 309. If X is connected, then all quotient spaces of X are connected.

Theorem 310. Let L be a space with order topology. If L satisfies least upper bound and density, then L is connected.

Theorem 311. For $x, y \in X$, $x \sim y$ iff there is a connected $A \subseteq X$ with $x, y \in A$, is an equivalence relation. The equivalence classes are the *connected components* of X.

Theorem 312. The connected components of X are all closed and connected.

Theorem 313. X is disconnected iff $\exists A \subset X$ with $A \neq X, \varnothing$ which is both open and closed.

Definition 168. Let X be a topological space. An *open covering* of X is a collection $C \subseteq \tau$ such that $X = \bigcup C$. X is *compact* iff every open covering of X has a finite subset covering X.

Theorem 314. If $f: X \to Y$ is continuous and X is compact, then $f(X)$ is compact.

Theorem 315. The discrete topology is compact iff it is finite. The trivial topology is always compact.

Theorem 316. Every finite space is compact.

Theorem 317. If X is compact and $A \subseteq X$ is closed, then A is compact.

Theorem 318. If X is Hausdorff and $A \subseteq X$ is compact, then A is closed.

Theorem 319. If $A, B \subseteq X$ are both compact, then $A \cap B$ and $A \cup B$ are compact.

Theorem 320. (Tychonoff) Every product of compact spaces is compact.

Theorem 321. If X is compact, then all quotient spaces of X are compact.

Theorem 322. An order topology satisfies least upper bound iff every closed interval is compact.

Definition 169. X is *locally compact* iff $\forall x \in X$, there is a compact $K \subseteq X$ and an open $U \subseteq X$ with $x \in U \subseteq K$.

Definition 170. X is *Bolzano-Weierstrass* iff every infinite subset of X has a limit point.

Definition 171. X is *sequentially compact* iff every sequence in X has a convergent sub-permutation.

Theorem 323. If X is compact then X is Bolzano-Weierstrass. If X is sequentially compact, then X is Bolzano-Weierstrass.

Definition 172. Let X be a set. A *metric* on X is a function $d\colon X \times X \to \mathbb{R}$ satisfying
1. (positive definite) $d(x,y) \geq 0$, and $d(x,y) = 0$ iff $x = y$.
2. (symmetric) $d(x,y) = d(y,x)$.
3. (triangle inequality) $d(x,y) \leq d(x,z) + d(z,y)$.

A *metric space* is a set with a metric.

Definition 173. Let X be a metric space, $x \in X$, $r > 0$. $b_r(x) \equiv \{y \in X \mid d(x,y) < r\}$.

Theorem 324. A metric space X has a unique topology generated from the base $B_d \equiv \{b_r(x) \mid x \in X,\ r > 0\}$.

Theorem 325. $\mathbb{R}$ is a metric space under $d(x,y) = |x - y|$. Its metric topology coincides with its order topology.

Definition 174. A metric space X is *bounded* iff $\exists M > 0$ such that $\forall x, y \in X$, $d(x,y) < M$.

Definition 175. Let $A, B \subseteq X$. $d(A,B) \equiv \inf\{d(x,y) \mid x \in A,\ y \in B\}$.

Definition 176. Let X, Y be metric spaces. A function $f\colon X \to Y$ is an *isometry* iff $\forall x, y \in X$, $d(x,y) = d\left(f(x), f(y)\right)$.

Theorem 326. Every isometry is injective and continuous.

Theorem 327. Every surjective isometry is a homeomorphism.

Theorem 328. In a metric space, a sequence $\{x_n\}$ converges to x iff $\lim d(x_n, x) = 0$.

Definition 177. In a metric space, a sequence $\{x_n\}$ is *Cauchy* iff $\forall \epsilon > 0$, $\exists N \in \mathbb{N}$ such that $\forall n, m > N$, $d(x_n, x_m) < \epsilon$.

Theorem 329. Every convergent sequence is Cauchy.

Theorem 330. If a sequence $\{x_n\}$ is Cauchy and has a sub-permutation converging to x, then $\{x_n\}$ converges to x.

Definition 178. A metric space is *complete* iff every Cauchy sequence converges.

Theorem 331. If X and Y are complete, then $X \times Y$ is complete.

Theorem 332. Let X be complete. Then $A \subseteq X$ is complete iff A is closed.

Theorem 333. $\mathbb{R}$ is complete.

Theorem 334. If X is complete and dense, then X is uncountable.

Definition 179. A metric space X is *totally bounded* iff $\forall \epsilon > 0$ there is a finite $\{x_i\} \subseteq X$ such that $X = \bigcup_i b_\epsilon(x_i)$.

Theorem 335. If X is totally bounded, then X is bounded.

Theorem 336. (Heine-Borel) A metric space is compact iff it is complete and totally bounded.

Definition 180. A topological space X is *T1* iff $\forall x \in X$, $\{x\}$ is closed.

Definition 181. A topological space X is *regular* iff $\forall x \in X$, $\forall F \subseteq X$ closed with $x \notin F$, $\exists U_1, U_2$ open such that $x \in U_1$, $F \subseteq U_2$, and $U_1 \cap U_2 = \varnothing$.

Definition 182. A topological space X is *normal* iff $\forall F_1, F_2 \subseteq X$ closed with $F_1 \cap F_2 = \varnothing$, $\exists U_1, U_2$ open such that $F_1 \subseteq U_1$, $F_2 \subseteq U_2$, and $U_1 \cap U_2 = \varnothing$.

Theorem 337. If X is T1 and normal, then X is regular.

Theorem 338. If X is T1 and regular, then X is Hausdorff.

Theorem 339. If X is Hausdorff, then X is T1.

Definition 183. A topological space X is *1st countable* iff $\forall x \in X$, there is a countable $B_x \subseteq \tau$ such that $\forall b \in B_x$, $x \in b$, and for every neighborhood U of x, $\exists b \in B_x$ with $b \subseteq U$.

Definition 184. A topological space is *2nd countable* iff it has a countable base.

Theorem 340. Every 2nd countable space is 1st countable.

Theorem 341. Every subspace of a T1 space is T1. Every product of T1 spaces is T1.

Theorem 342. Every subspace of a Hausdorff space is Hausdorff. Every product of Hausdorff spaces is Hausdorff.

Theorem 343. Every subspace of a regular space is regular. Every product of regular spaces is regular.

Theorem 344. Every subspace of a 1st countable space is 1st countable. Every countable product of 1st countable spaces is 1st countable.

Theorem 345. Every subspace of a 2nd countable space is 2nd countable. Every countable product of 2nd countable spaces is 2nd countable.

Theorem 346. Every subspace of a metric space is metric. Every countable product of metric spaces is metric.

Theorem 347. If $\prod X_\alpha$ is T1, Hausdorff, regular, normal, 1st countable, 2nd countable, metric, compact, or connected, then so is X_α $\forall \alpha$.

Theorem 348. Every compact Hausdorff space is normal.

Theorem 349. Every locally compact Hausdorff space is regular.

Theorem 350. Every compact metric space is 2nd countable.

Theorem 351. Every metric space is 1st countable, normal, and T1.

Theorem 352. If X is 2nd countable, regular, and T1, then X is metric.

Theorem 353. A finite space is metric iff it is discrete.

Theorem 354. Every discrete space is metric.

Theorem 355. Every order topology is regular and T1.

Theorem 356. Let X be 1st countable, $A \subseteq X$. If $x \in \bar{A}$ then there exists a sequence $\{a_n\} \subseteq A$ converging to X.

Theorem 357. Let X be 1st countable, $f\colon X \to Y$. If $\forall x \in X$ and for every sequence $\{x_n\}$ converging to x, $\{f(x_n)\}$ converges to $f(x)$, then f is continuous.

Theorem 358. If X is 1st countable and Bolzano-Weierstrass, then X is sequentially compact.

Theorem 359. If X is metric and sequentially compact, then X is compact.

Definition 185. A *compactification* of X is a compact Hausdorff space Y such that $X \subseteq Y$ is dense in Y. X is *compactifiable* iff it has a compactification.

Theorem 360. Every normal T1 space is compactifiable.

Theorem 361. If a space is compactifiable, then it is regular and T1.

Theorem 362. If a space is locally compact and Hausdorff, then it is compactifiable.

Definition 186. A compactification $\beta(X)$ of X is *Stone-Čech* iff for every compact Hausdorff K and every continuous $f\colon X \to K$, there is a unique $f'\colon \beta(X) \to K$ with $f(x) = f'(x)\ \forall x \in X$.

Theorem 363. Every compactifiable space has a Stone-Čech compactification.

Theorem 364. Any two Stone-Čech compactifications of X are homeomorphic, with identity on X.

Theorem 365. If X is compactifiable, then every compactification of X is a quotient space of $\beta(X)$ preserving X.

Definition 187. A *topological group* is a group which is a T1 topological space, such that group composition and inverse are continuous.

Theorem 366. Every topological group is regular.

Theorem 367. Every topological group is homogeneous.

Theorem 368. Every product of topological groups is a topological group.

Theorem 369. Every subgroup of a topological group is a topological group.

Theorem 370. Let G be a topological group. If $H \leq G$, then $\bar{H} \leq G$.

Theorem 371. Let G be a topological group. If $H \trianglelefteq G$, then $\bar{H} \trianglelefteq G$.

Theorem 372. G/H is a topological group iff H is closed.

Chapter 4

Analysis et al.

4.1 Measure Theory

Definition 188. Let X be a set and S a δ-ring on X such that $\bigcup S = X$. A *measure* is a function $\mu \colon S \to \mathbb{R}$ satisfying
1. $\forall E \in S,\ \mu(E) \geq 0$
2. $\mu(\varnothing) = 0$
3. $\forall \{E_i\} \subseteq S$, countable, such that $E_i \cap E_j = \varnothing$,

$$\mu\left(\bigcup_i E_i\right) = \sum \mu(E_i)$$

Theorem 373. If $E \subseteq F$ then $\mu(E) \leq \mu(F)$.

Theorem 374. $\mu(E \cup F) = \mu(E) + \mu(F) - \mu(E \cap F)$

Theorem 375. If $\{E_i\} \subseteq S$ such that $\forall i,\ E_i \subseteq E_{i+1}$, then

$$\mu(\lim E_i) = \lim \mu(E_i)$$

If $\{E_i\} \subseteq S$ such that $\forall i,\ E_{i+1} \subseteq E_i$, then

$$\mu(\lim E_i) = \lim \mu(E_i)$$

Definition 189. A measure is *complete* iff $\forall E \in S$, if $\mu(E) = 0$ and $F \subseteq E$, then $F \in S$.

Theorem 376. Let μ be a measure on S. Let $\bar{S} \equiv \{E + N \mid E \in S, \; N \subseteq F, \; \mu(F) = 0\}$. Let $\bar{\mu}(E + N) \equiv \mu(E)$. Then $\bar{S}$ is a δ-ring, and $\bar{\mu}$ is a complete measure.

Theorem 377. Let μ, ν be measures on S. Let $\alpha \geq 0$. Then $\mu + \nu$ and $\alpha\mu$ are measures on S.

Definition 190. Measures μ and ν are *equivalent* $(\mu \equiv \nu)$ iff $\forall E \in S$,

$$\mu(E) = 0 \Leftrightarrow \nu(E) = 0$$

Theorem 378. $\mu \equiv \nu$ is an equivalence relation.

Theorem 379. $\{E \in S \mid \mu(E) = 0\}$ is an ideal subring of S.

Definition 191. Let S_X, S_Y be δ-rings of X, Y respectively. $S_{X \times Y}$ is the δ-ring on $X \times Y$ generated by $\{E_X \times E_Y \mid E_X \in S_X, \; E_Y \in S_Y\}$

Theorem 380. Let μ, ν be measures on S_X, S_Y respectively. There is a unique measure $\mu \times \nu$ on $S_{X \times Y}$ such that $\forall E_X \in S_X, \; E_Y \in S_Y$,

$$(\mu \times \nu)(E_X \times E_Y) = \mu(E_X) \cdot \nu(E_Y)$$

Theorem 381. $\alpha(\mu \times \nu) = (\alpha\mu) \times \nu = \mu \times (\alpha\nu)$,
and $(\mu_1 + \mu_2) \times \nu = (\mu_1 \times \nu) + (\mu_2 \times \nu)$

Theorem 382. Let μ be a measure on (X, S). Let $f\colon X \to Y$. Then $f(S) \equiv \{E \subseteq Y \mid f^{-1}(E) \in S\}$ is a δ-ring on $f(X)$, and $f(\mu)(E) \equiv \mu(f^{-1}(E))$ is a measure on $f(S)$.

Definition 192. Let μ be a measure on S. Let $E \in S$. Let $f\colon E \to \mathbb{R}$ be bounded. Let P be a finite measurable partition of E.

$$U(f, P) \equiv \sum_{C_i \in P} \mu(C_i) \sup_{x \in C_i} f(x)$$

$$L(f, P) \equiv \sum_{C_i \in P} \mu(C_i) \inf_{x \in C_i} f(x)$$

Definition 193. $f\colon X \to \mathbb{R}$ is *integrable* on E iff

$$\inf_P U(f, P) = \sup_P L(f, P) \equiv \int_E f \ d\mu$$

This definition of integration is not the most general or the most useful in all contexts. It can be extended in various ways, which are beyond the scope of this book.

Theorem 383. If $\mu(E) = 0$, then $\int_E f \ d\mu = 0$.

Theorem 384. $\int_E 1 \ d\mu = \mu(E)$

Theorem 385. $\int_E f \circ g \ d\mu = \int_{g(E)} f \ dg(\mu)$

Theorem 386. $\int_E \alpha f + \beta g \ d\mu = \alpha \int_E f \ d\mu + \beta \int_E g \ d\mu$

Theorem 387. $\int_E f \ d(\alpha\mu + \beta\nu) = \alpha \int_E f \ d\mu + \beta \int_E f \ d\nu$

Theorem 388. If $\mu(E_1 \cap E_2) = 0$, then $\int_{E_1 \cup E_2} f \ d\mu = \int_{E_1} f \ d\mu + \int_{E_2} f \ d\mu$.

Theorem 389. If $f \leq g$, then $\int_E f \ d\mu \leq \int_E g \ d\mu$.

Theorem 390. $\left| \int_E f \ d\mu \right| \leq \int_E |f| \ d\mu$

Theorem 391. (Fubini) Let $X \times Y$ have measure $\mu \times \nu$. Let $h\colon X \times Y \to \mathbb{R}$, $A \in S_X$, $B \in S_Y$. Then

$$\int_{A \times B} h \ d(\mu \times \nu) = \int_A \int_B h \ d\nu \ d\mu = \int_B \int_A h \ d\mu \ d\nu$$

If $h(x, y) = f(x)g(y)$, then

$$\int_{A \times B} h \ d(\mu \times \nu) = \int_A f \ d\mu \cdot \int_B g \ d\nu$$

Theorem 392. (Radon-Nikodym) Let μ, ν be measures on (X, S) such that $\mu \equiv \nu$. Then $\exists f \colon X \to X$ such that $\forall E \in S$,

$$\nu(E) = \int_E f \, d\mu$$

denoted

$$f \equiv \frac{d\nu}{d\mu}$$

Theorem 393.
$$\frac{d}{d\mu}(\alpha\nu + \beta\lambda) = \alpha\frac{d\nu}{d\mu} + \beta\frac{d\lambda}{d\mu}$$

Theorem 394.

$$\frac{d\nu}{d\lambda} = \frac{d\nu}{d\mu}\frac{d\mu}{d\lambda} \qquad \frac{d\nu}{d\mu} = \left(\frac{d\mu}{d\nu}\right)^{-1}$$

Theorem 395.
$$\int_E f \, d\mu = \int_E f \frac{d\mu}{d\lambda} \, d\lambda$$

Theorem 396.
$$\frac{d(\mu_1 \times \mu_2)}{d(\nu_1 \times \nu_2)} = \frac{d\mu_1}{d\nu_1}\frac{d\mu_2}{d\nu_2}$$

Definition 194. Let X be a topological space. The *Borel* δ-ring B on X is the δ-ring generated by the compact subsets of X. A measure on B is a *Borel* measure.

Theorem 397. If $E \in B$ then so are $\mathring{E}$ and $\bar{E}$.

Theorem 398. Let μ be a Borel measure, let E be compact. If f is continuous on E, then it is integrable on E over μ.

Definition 195. Let μ be a Borel measure on X. The *support* of μ is $\operatorname{supp}\mu \subseteq X$ such that $x \in \operatorname{supp}\mu$ iff $\forall$ open $U \ni x$, $\mu(U) > 0$.

Theorem 399. $\operatorname{supp}\mu$ is closed.

Theorem 400. If U is open and $U \cap \mathrm{supp}\mu \neq \varnothing$, then $\mu(U) > 0$.

Theorem 401. $\mathrm{supp}\mu = X$ iff $\mu(U) > 0 \ \forall$ open $U \neq \varnothing$.

Theorem 402. $\mathrm{supp}\mu^c = \bigcup\{U \mid U \text{ is open and } \mu(U) = 0\}$

Theorem 403. Let X be compact Hausdorff. If $\exists E$ such that $\mu(E) > 0$, then $\mathrm{supp}\mu \neq \varnothing$.

Definition 196. Let G be a locally compact topological group. A Borel measure μ on G is a *Haar* measure iff
1. $\forall U \in B$ open and $U \neq \varnothing$, $\mu(U) > 0$
2. $\forall g \in G, \ \forall E \in B, \ \mu(gE) = \mu(E)$

Theorem 404. (Haar) Every locally compact topological group has a Haar measure μ, and a measure ν is Haar iff $\exists \alpha > 0$ such that $\nu = \alpha\mu$.

Theorem 405. If μ, ν are Haar measures on G, H respectively, then $\mu \times \nu$ is a Haar measure on $G \times H$.

Definition 197. Let G be a group of automorphisms of (X, S). A measure μ on (X, S) is *G-quasi-invariant* iff $\forall g \in G, \ g(\mu) \equiv \mu$.

Theorem 406. Let B be the Borel δ-ring on X, and let G be a locally compact, transitive group of homeomorphisms of X. $\exists$ measure μ on B such that
1. $\forall U \in B$ open and $U \neq \varnothing$, $\mu(U) > 0$
2. μ is G-quasi-invariant

If ν satisfies the above, then $\nu \equiv \mu$.

Theorem 407. If μ is G-quasi-invariant, then $\mathrm{supp}\mu$ is invariant under G.

Theorem 408. If μ is G-quasi-invariant and G is transitive, then either $\mathrm{supp}\mu = \varnothing$ or $\mathrm{supp}\mu = X$.

Definition 198. Let G be a locally compact group of homeomorphisms of X, with Haar measure μ. For $f\colon X \to \mathbb{R}$, define $f^\flat\colon X/G \to \mathbb{R}$ by

$$f^\flat(G(x)) \equiv \int_G f(g(x))\ d\mu$$

Theorem 409. Given a Borel measure λ on X/G, there is a unique measure $\lambda^\sharp$ on X such that $\forall f$,

$$\int_{X/G} f^\flat\ d\lambda = \int_X f\ d\lambda^\sharp$$

Theorem 410. $\forall g \in G,\ g(\lambda^\sharp) = \lambda^\sharp$

Theorem 411. $\lambda \equiv \lambda'$ iff $\lambda^\sharp \equiv \lambda'^\sharp$

Theorem 412. $\lambda^\sharp(X) = \mu(G) \cdot \lambda(X/G)$

4.2 Functional Analysis

Definition 199. A *topological vector space* (TVS) is a vector space with T1 topology such that addition and scalar multiplication are continuous.

Theorem 413. Every topological vector space is regular.

Theorem 414. Any product of TVSs is a TVS.

Theorem 415. Let V be a TVS. If $W \leq V$ then $\overline{W} \leq V$.

Theorem 416. Let V be a TVS. V/W is a TVS iff W is closed.

Theorem 417. Let $L \in \mathcal{L}(V)$ be continuous. Then $\ker L$ is closed.

Theorem 418. Let $W \leq V$. If W or V/W is finite dimensional, then W is closed.

Theorem 419. Let $\{a_n\}, \{b_n\}$ be sequences in a TVS. The following hold:

$$\lim(a_n + b_n) = \lim a_n + \lim b_n$$
$$\lim \alpha a_n = \alpha \lim a_n$$
$$\sum(a_n + b_n) = \sum a_n + \sum b_n$$
$$\sum \alpha a_n = \alpha \sum a_n$$

Theorem 420. A TVS is locally compact iff it is finite dimensional.

Theorem 421. If V is finite dimensional, there is a unique topology on V such that it is a TVS. It is given by the product topology on F^n.

Definition 200. A collection $\{e_\mu\} \subseteq V$ is a (Schauder/topological) *basis* iff for each $v \in V$, there is a unique sequence $\{\alpha^\mu\} \subseteq F$ such that

$$v = \sum \alpha^\mu e_\mu$$

Definition 201. A collection $\{e_\mu\} \subseteq V$ is *total* iff span$\{e_\mu\}$ is dense in V.

Theorem 422. $\{e_\mu\}$ is total iff $\forall v \in V$, $\exists \{\alpha^\mu\} \subseteq F$ such that

$$v = \sum \alpha^\mu e_\mu$$

Theorem 423. Every basis is total and linearly independent.

Theorem 424. For $i \in I$, define $e_i \colon I \to F$ as

$$e_i(j) \equiv \begin{cases} 1 & i = j \\ 0 & i \neq j \end{cases}$$

Then $\{e_i\}$ is a topological basis for the full F^I.

Definition 202. Let V be a TVS over F. Its *topological dual* V^* is the space of continuous linear functionals $\omega \colon V \to F$.

Theorem 425. If V, W are finite dimensional TVSs, then every linear map $L\colon V \to W$ is continuous.

Definition 203. Let $\{e_\mu\}$ be a basis. The *coordinate functionals* are the linear maps $\{e^\mu\colon V \to F\}$ such that $e^\mu(e_\nu) = \delta^\mu{}_\nu$.

Theorem 426. $\{e^\mu\}$ are linearly independent.

Theorem 427. Let $\{V_i\}$ be TVSs. $\left(\prod V_i\right)^* = \bigoplus (V_i^*)$ and $\left(\bigoplus V_i\right)^* = \prod (V_i^*)$.

Definition 204. Let V be a vector space over $\mathbb{R}$ or $\mathbb{C}$. A *norm* on V is a function $\|\cdot\|\colon V \to \mathbb{R}$ satisfying

1. $\|v\| \geq 0$, and $\|v\| = 0$ iff $v = 0$
2. $\|\alpha v\| = |\alpha| \|v\|$
3. $\|v + w\| \leq \|v\| + \|w\|$

Theorem 428. For any inner product, $\sqrt{\langle v, v \rangle}$ is a norm.

Theorem 429. If $\|\cdot\|$ is a norm and $\alpha > 0$, then $\alpha\|\cdot\|$ is a norm.

Theorem 430. For any norm, $d(v, w) \equiv \|v - w\|$ is a metric.

Theorem 431. (Mazur-Ulam) Every surjective isometry of a real normed vector space is affine.

Theorem 432. Every normed vector space is a TVS in the associated metric topology, and the norm is continuous.

Theorem 433. Let V, W be normed spaces. A linear map $L\colon V \to W$ is continuous iff $\exists C > 0$ such that $\|L(v)\| \leq C\|v\| \ \forall v \in V$.

Theorem 434. Let V, W be normed spaces. The space $B(V, W) \leq \mathcal{L}(V, W)$ of continuous linear maps is a normed space with

$$\|L\| \equiv \sup_v \left(\frac{\|L(v)\|}{\|v\|} \right)$$

Theorem 435. $\|L^T\| = \|L\|$

Theorem 436. Let V be normed. Then the map $i\colon V \to V^{**}$ such that $i(v)(\omega) = \omega(v)$ is a linear isometry.

Theorem 437. Let V be normed. If $\omega(v) = 0 \ \forall \omega \in V^*$, then $V = 0$.

Theorem 438. Let V be normed. $\forall p \geq 1$, $\|v\|_p \equiv \left(\sum \|v_i\|^p\right)^{1/p}$ is a norm on V^n.

Theorem 439. Let V be normed. There exists a complete normed space $\bar{V}$ such that V is a dense subspace of $\bar{V}$.

Definition 205. $\bar{V}$ is a *completion* of V.

Theorem 440. Any two completions of V are isometrically isomorphic, with identity on V.

Theorem 441. Let C be the space of all Cauchy sequences in V. Let $N \leq C$ consist of all sequences converging to zero. Then C/N is a completion of V.

Definition 206. A complete normed space is a *Banach* space.

Theorem 442. Every finite-dimensional normed space is Banach.

Definition 207. A summed sequence $\sum a_n$ in a normed space converges *absolutely* iff $\sum \|a_n\|$ converges.

Theorem 443. A normed space is Banach iff every absolutely convergent summed sequence is convergent.

Theorem 444. If V is Banach and $\sum a_n$ in V converges absolutely, then for every permutation σ of $\mathbb{N}$,

$$\sum a_{\sigma(n)} = \sum a_n$$

Theorem 445. If V is Banach then V^n is Banach with $\|\cdot\|_p$.

Theorem 446. If V is normed and W is Banach, then $B(V, W)$ is Banach.

Theorem 447. A Banach space is *reflexive* iff the map $i\colon V \to V^{**}$ such that $i(v)(\omega) = \omega(v)$ is an isomorphism.

Theorem 448. V is reflexive iff V^* is reflexive.

Theorem 449. If $W \leq V$ is closed and V is reflexive then W and V/W are reflexive.

Theorem 450. If V is Banach and $\{e_\mu\}$ is a basis for V, then the coordinate functionals are continuous.

Theorem 451. If V is reflexive and $\{e_\mu\}$ is a basis for V, then the coordinate functionals are a basis for V^*.

Definition 208. Let (X, μ) be a measure space. Let V be Banach. Let $p \geq 1$. Define the spaces

$$I_p(\mu, V) \equiv \{f\colon X \to V \mid \|f\|^p \text{ is integrable on } X\}$$

$$N(\mu, V) \equiv \{f\colon X \to V \mid \mu\left(f^{-1}(V \setminus \{0\})\right) = 0\}$$

$$L_p(\mu, V) \equiv I_p(\mu, V)/N(\mu, V)$$

Theorem 452. $L_p(\mu, V)$ is a Banach space with

$$\|f\|_p \equiv \left(\int_X \|f\|^p \, d\mu\right)^{1/p}$$

Theorem 453. $L_p(\mu, V)^* = L_q(\mu, V^*)$, where $\frac{1}{p} + \frac{1}{q} = 1$, and

$$f(g) = \int_X f(g) \, d\mu$$

Theorem 454. If V is reflexive and $p \neq 1$, then $L_p(\mu, V)$ is reflexive.

Definition 209. Let X be a set, let V be Banach. Define $L_\infty(X, V)$ to be the space of bounded functions $f\colon X \to V$.

Theorem 455. $L_\infty(X, V)$ is Banach with $\|f\|_\infty \equiv \sup \|f(x)\|$.

Theorem 456. Let μ be a measure, let $p \leq q$.
If $\exists C \in \mathbb{R}$ such that $\forall E$, $\mu(E) < C$, then $L_q(\mu) \leq L_p(\mu)$.
If $\exists C > 0$ such that $\forall E$, $\mu(E) > C$, then $L_p(\mu) \leq L_q(\mu)$.

Theorem 457. $L_p(A \times B) = L_p(A) \otimes L_p(B)$

Definition 210. Let V, W be Banach. A function $f: V \to W$ is *differentiable* at $v \in V$ iff there exists a map $Df(v) \in B(V, W)$ such that

$$\lim_{h \to 0} \frac{f(v + h) - f(v) - Df(v)(h)}{\|h\|} = 0$$

Theorem 458. If f is differentiable at v, then $Df(v)$ is unique.

Definition 211. Let f be differentiable at all $v \in V$. The *derivative* of f is the function $V \to B(V, W)$ given by $Df(v)$.

Theorem 459. If f is differentiable then it is continuous.

Facts:

$D(\alpha f + \beta g) = \alpha Df + \beta Dg$
$D(f \otimes g) = Df \otimes g + f \otimes Dg$
$D(f \circ g) = Df \cdot Dg$
$D(f^{-1}) = Df^{-1}$
$Df(v) = 0 \ \forall v$ iff f is constant
If $L \in B(V, W)$, then $DL = L$
$D_a D_b f = D_b D_a f$

Theorem 460. Let V, W be Banach and let W have a basis. Let $f: V \to W$ be differentiable. Then the components $f^\mu \equiv e^\mu(f)$ are differentiable, and

$$Df = \sum e_\mu \otimes Df^\mu$$

Definition 212. A function $f: V \to W$ is *smooth* iff $\forall n \in \mathbb{N}$, $D^n f$ is differentiable. The space of smooth functions is denoted $C_\infty(V, W)$.

Theorem 461. A function $f\colon \mathbb{C} \to \mathbb{C}$ is smooth iff it is differentiable.

Theorem 462. $\{z^n \mid n \in \mathbb{N}\}$ is a basis for $C_\infty(\mathbb{C})$.

Theorem 463. Let V, W be finite dimensional and Banach over $\mathbb{C}$, let $f\colon V \to W$ be smooth. Then

$$f(v) = \sum \frac{1}{n!} D^n f(0) \left(\otimes^n v\right)$$

Definition 213. Define $\exp\colon \mathbb{C} \to \mathbb{C}$ as $\exp(z) \equiv \sum \frac{z^n}{n!}$

Facts:

$\exp(z + w) = \exp(z)\exp(w)$
$\exp(0) = 1$
$\forall z,\ \exp(z) \neq 0$
$\forall n \in \mathbb{Z},\ \exp(z)^n = \exp(nz)$
$D(\exp) = \exp$

Definition 214. Define $\exp\colon B(V) \to B(V)$ as $\exp(L) \equiv \sum \frac{L^n}{n!}$

Facts:

$\exp((\alpha + \beta)L) = \exp(\alpha L)\exp(\beta L)$
$\exp(0) = \mathbf{1}$
$\forall L,\ \exp(L) \neq 0$
$\forall n \in \mathbb{Z},\ \exp(L)^n = \exp(nL)$
$\exp(TLT^{-1}) = T\exp(L)T^{-1}$
$\det(\exp L) = \exp(\mathrm{Tr}L)$
$D^a{}_b(\exp L^c{}_d) = \delta^c{}_b \exp L^a{}_d$

Definition 215. Let (X, S, μ) be a measure space and let $E \in S$. Let V be Banach. A function $f\colon X \to V$ is *integrable* on E iff $\exists I \in V$ such that $\forall \omega \in V^*,\ \omega \circ f$ is integrable on E and

$$\int_E \omega \circ f \ d\mu = \omega(I)$$

Define $\int_E f \ d\mu \equiv I$.

Theorem 464. $\int (\alpha f + \beta g) = \alpha \int f + \beta \int g$

Theorem 465. For $L \in B(V, W)$, $\int L(f) = L\left(\int f\right)$

Theorem 466. Let $v \in V$, $\phi \colon X \to \mathbb{R}$. Then $\int \phi v = v \int \phi$.

Theorem 467. $\left\| \int f \right\| \le \int \|f\|$

Theorem 468. Let V be Banach with a basis. Let $f \colon X \to V$ be integrable. Then the components f^μ are integrable and

$$\int f \, d\mu = \sum \int f^\mu \, d\mu \, e_\mu$$

Definition 216. A *Hilbert* space is a complete inner product space.

Theorem 469. In a Hilbert space, every orthogonal total set is a basis.

Theorem 470. Every Hilbert space has an orthonormal basis.

Theorem 471. All orthonormal bases have the same cardinality.

Theorem 472. Let (X, μ) be a measure space. Let H be a Hilbert space. $L_2(\mu, H)$ is a Hilbert space with inner product

$$\langle\langle f, g \rangle\rangle \equiv \int_X \langle f, g \rangle \, d\mu$$

Theorem 473. (Riesz) In a Hilbert space, the adjoint map $\dagger \colon H \to H^*$ is an isometric antilinear isomorphism.

Theorem 474. Let H, H' be Hilbert and let $L \in B(H, H')$. There exists a unique map $L^\dagger \in B(H', H)$ such that $\forall v \in H$, $\forall w \in H'$,

$$\langle w, L(v) \rangle = \langle L^\dagger(w), v \rangle$$

Theorem 475. Adjoint $\dagger \colon B(H, H') \to B(H', H)$ is an antilinear isomorphism.

Definition 217. Let H be Hilbert. Let $V \leq H$. Define $V^\perp \equiv \{w \in H \mid \langle v, w \rangle = 0 \ \forall v \in V\}$.

Facts:

$V^\perp \leq H$
$V^\perp$ is closed
$V^{\perp\perp} = \bar{V}$
$V \leq W$ iff $W^\perp \leq V^\perp$
If $V \leq W \leq \bar{V}$ then $W^\perp = V^\perp$
$(V + W)^\perp = V^\perp \cap W^\perp$

Theorem 476. If V is closed, then $H = V \oplus V^\perp$ orthogonally.

Theorem 477. Let $L \in B(H)$ be Hermitian. Then $H = \ker L \oplus \overline{L(H)}$ orthogonally.

Theorem 478. Let H be Hilbert. For every $V \leq H$ closed, there exists a unique *projection* $P_V \in B(H)$ such that
1. $P_V = P_V^\dagger$
2. $P_V^2 = P_V$
3. $P_V(H) = V$

Theorem 479. P_V is identity on V, and $\|P_V\| = 1$.

Theorem 480. $\ker P_V = V^\perp$

Theorem 481. $P_V + P_{V^\perp} = \mathbf{1}$

Theorem 482. $P_V P_W = 0$ iff V and W are orthogonal.

Definition 218. $L \in B(H)$ is *Hilbert-Schmidt* iff $\operatorname{Tr}(L^\dagger L)$ exists.

Theorem 483. The space of Hilbert-Schmidt maps is a Hilbert space under the inner product $\operatorname{Tr}(A^\dagger B)$.

Theorem 484. (Spectral) If L is Hilbert-Schmidt and Hermitian, then $H = \bigoplus_\lambda H_L(\lambda)$ orthogonally, and

$$L = \sum_\lambda \lambda v_\lambda \otimes v_\lambda^\dagger$$

where λ ranges over the eigenvalues of L, and $\{v_\lambda\}$ is an orthonormal basis of eigenvectors of L.

Definition 219. A *Banach algebra* is an algebra A with 1 which is a Banach space, such that
1. $\|xy\| \le \|x\|\|y\| \; \forall x, y \in A$
2. $\|1\| = 1$

Theorem 485. If A is a Banach algebra, then $A^\times$ is a topological group.

Theorem 486. Let V be Banach, let A be a Banach algebra. Let $f, g \colon V \to A$. Then $D(fg) = D(f)g + fD(g)$.

Definition 220. For x in a Banach algebra, the *spectrum* of x is

$$\sigma(x) \equiv \{\lambda \in \mathbb{C} \mid x - \lambda 1 \text{ is not invertible}\}$$

Theorem 487. If A is a Banach algebra over $\mathbb{C}$, then $\forall x \in A$, $\sigma(x)$ is non-empty and compact.

Definition 221. A **-algebra* is an algebra A with 1 over $\mathbb{C}$ with a map $* \colon A \to A$, such that $\forall x, y \in A$, $\forall \alpha \in \mathbb{C}$,
1. $(x + y)^* = x^* + y^*$
2. $(xy)^* = y^* x^*$
3. $x^{**} = x$
4. $(\alpha x)^* = \bar{\alpha} x^*$

Theorem 488. $1^* = 1$

Definition 222. Let A be a *-algebra.

$$A_R \equiv \{x \in A \mid x^* = x\}$$
$$A_I \equiv \{x \in A \mid x^* = -x\}$$

Theorem 489. $A = A_R \oplus A_I$ over $\mathbb{R}$, and $A_I = iA_R$.

Theorem 490. Let V be Banach, let A be a *-algebra, let $f\colon V \to A$. Then $D(f^*) = (Df)^*$.

Theorem 491. Let X be a measure space, let A be a *-algebra, let $f\colon X \to A$. Then $\int f^* = \left(\int f\right)^*$.

Definition 223. A *state* on *-algebra A is a linear functional $E \in A^*$ such that
1. $E(1) = 1$
2. $E(x^*x) \geq 0 \ \forall x \in A$

Theorem 492. $E(x^*) = \overline{E(x)}$

Definition 224. A state is *pure* iff it is not a positive linear combination of states.

Theorem 493. If $\{E_i\}$ are states and $\{\lambda_i\} \subseteq \mathbb{R}$ such that $\sum \lambda_i = 1$ and $\lambda_i \geq 0 \ \forall i$, then $\sum \lambda_i E_i$ is a state.

Definition 225. A *C*-algebra* is a Banach *-algebra such that $\forall x$, $\|x^*x\| = \|x\|\|x^*\|$.

Theorem 494. In a C*-algebra, $\|x\| = \|x^*\|$.

Theorem 495. In a C*-algebra, $\|x\|^2 = \sup |\sigma(x^*x)|$.

Theorem 496. In a C*-algebra, if $x^* = x$, then $\sigma(x) \subseteq \mathbb{R}$, and $\|x\| = \sup |\sigma(x)|$.

Theorem 497. Let E be a state on a C*-algebra. $\|E\| = 1$.

Theorem 498. Let H be a Hilbert space. $B(H)$ is a C*-algebra under adjoint.

Theorem 499. If A is a Banach or C*-algebra, then so is $L_\infty(X, A)$ under pointwise multiplication.

Theorem 500. If A is a Banach algebra, then under pointwise multiplication,

$$L_p(\mu, A) \cdot L_q(\mu, A) = L_r(\mu, A)$$

where $\frac{1}{p} + \frac{1}{q} = \frac{1}{r}$.

Definition 226. Let A be a C*-algebra, let H be a Hilbert space. A *representation* of A on H is a *-homomorphism $R \colon A \to B(H)$.

Definition 227. Let R_1, R_2 be representations of A on H_1, H_2 respectively. Their *direct sum* is the representation $R_1 \oplus R_2$ on $H_1 \oplus H_2$ given by

$$(R_1 \oplus R_2)(x)(v_1 \oplus v_2) \equiv R_1(x)(v_1) \oplus R_2(x)(v_2)$$

Theorem 501. Let R be a representation of A on H. If there exists a closed, proper, nontrivial subspace $V < H$ such that $R(A)(V) \subseteq V$, then the restrictions $R_V, R_{V^\perp}$ of R to V and $V^\perp$ respectively, are representations of A, and $R = R_V \oplus R_{V^\perp}$.

Definition 228. A representation R of A on H is *irreducible* iff it is not a nontrivial direct sum of representations.

Definition 229. Let R be a representation of A on H. $v \in H$ is *cyclic* iff $R(A)(v)$ is dense in H.

Theorem 502. Let R be a representation of A on H. Let $v \in H$ such that $\|v\| = 1$. Then $E_v(x) \equiv \langle v, R(x)v \rangle$ is a state.

Theorem 503. Let v be cyclic in R. Then R is irreducible iff E_v is pure.

Theorem 504. R is irreducible iff $\forall v \neq 0$, v is cyclic.

Definition 230. Representations R_1, R_2 on H_1, H_2 respectively are *unitarily equivalent* iff there is a unitary map $U \colon H_1 \to H_2$ such that $\forall x \in A$, $R_2(x) = U R_1(x) U^{-1}$.

Theorem 505. Let v, v' be cyclic vectors in R, R' respectively, with $\|v\| = \|v'\| = 1$. If $E_v = E_{v'}$, then R and R' are unitarily equivalent.

Theorem 506. (Gelfand-Naimark-Segal) Let E be a state on A. Make the following definitions:

$$\langle x, y \rangle \equiv E(x^*y)$$

$$I \equiv \{x \in A \mid \langle x, x \rangle = 0\}$$

$$H \equiv A/I$$

For $x \in A$, let $|x\rangle$ denote the element of H corresponding to x. Define $R_E \colon A \to B(H)$ by

$$R_E(x)|y\rangle \equiv |xy\rangle$$

Then H is a Hilbert space under $\langle x, y \rangle$, R_E is a representation, $|1\rangle$ is cyclic, and

$$E(x) = \langle 1|R_E(x)|1\rangle$$

Theorem 507. Let A be a normed *-algebra, let R be a representation, let E be a state. For $x \in A$,

$$\sup_R \|R(x)\| = \sup_E \sqrt{E(x^*x)} = \sup \sqrt{|\sigma(x^*x)|}$$

Denote this quantity $\|x\|_C$. Let $I \equiv \{x \mid \|x\|_C = 0\}$. Then $C^*(A) \equiv \overline{A/I}$ is a C*-algebra with norm $\|x\|_C$, called the *enveloping* C*-algebra of A.

4.3 Differential Geometry

Definition 231. A *manifold* is a topological space M, 2nd countable Hausdorff, such that every point has a neighborhood homeomorphic to an open subset of $\mathbb{R}^n$. $\dim M \equiv n$.

Theorem 508. Every manifold has normal topology.

Theorem 509. If M, N are manifolds with dimensions m, n, then $M \times N$ is a manifold with dimension $m + n$.

Definition 232. A *smooth structure* on M is an open covering $\{U_\alpha\}$ of M, and homeomorphisms $\psi_\alpha\colon U_\alpha \to \mathbb{R}^n$ such that $\forall \alpha, \beta,\ \psi_\alpha \circ \psi_\beta^{-1}$ is smooth.

Definition 233. A function $f\colon M \to N$ between manifolds with smooth structure is *smooth* iff $\forall \alpha, \beta,\ \psi_\alpha \circ f \circ \psi_\beta^{-1}$ is smooth.

Definition 234. A function $f\colon M \to N$ is a *diffeomorphism* iff it is homeomorphic, smooth, and with a smooth inverse.

Theorem 510. On any set of manifolds, diffeomorphism is an equivalence relation.

Theorem 511. The set of diffeomorphisms $M \to M$, denoted $\mathrm{Diff}(M)$, is a group under composition of functions.

Theorem 512. Every manifold of dimension n is diffeomorphic to a submanifold of $\mathbb{R}^{2n+1}$.

Theorem 513. Let $f\colon M \to N$ be smooth. M is diffeomorphic to the manifold $\{(p, f(p)) \mid p \in M\}$.

Definition 235. Let M be a smooth manifold. $\mathcal{F}_M \equiv \{f\colon M \to \mathbb{R} \text{ smooth}\}$

Theorem 514. $\mathcal{F}_M$ is a commutative algebra with 1 under pointwise multiplication.

Theorem 515. A map $\delta \in \mathcal{F}_M^*$ is an algebra homomorphism $\mathcal{F}_M \to \mathbb{R}$ iff $\exists p \in M$ such that $\forall f \in \mathcal{F}_M,\ \delta(f) = f(p)$.

Theorem 516. $\mathcal{F}_M \simeq \mathcal{F}_N$ iff M is diffeomorphic to N.

Theorem 517. If $\phi\colon M \to N$ is a diffeomorphism, then $I_\phi\colon \mathcal{F}_N \to \mathcal{F}_M$ given by $I_\phi(f) = f \circ \phi$ is an isomorphism, and every isomorphism $I\colon \mathcal{F}_N \to \mathcal{F}_M$ is given by $I(f) = f \circ \phi$ for some diffeomorphism ϕ.

Theorem 518. $\mathrm{Aut}(\mathcal{F}_M) \simeq \mathrm{Diff}(M)$

Theorem 519. If N is a submanifold of M, then $\{f \in \mathcal{F}_M \mid f(N) = 0\} \trianglelefteq \mathcal{F}_M$, and $\mathcal{F}_M / \{f \in \mathcal{F}_M \mid f(N) = 0\} \simeq \mathcal{F}_N$.

Definition 236. Let $p \in M$. A *tangent vector* at p is a linear map $v \colon \mathcal{F}_M \to \mathbb{R}$ such that

$$v(fg) = f(p)v(g) + g(p)v(f)$$

Theorem 520. The space of tangent vectors at p, denoted TM_p, is a vector space, and $\dim TM_p = \dim M$.

Theorem 521. If N is a submanifold of M, then $TN_p \leq TM_p$.

Definition 237. Let U be an open subset of M. A *local coordinate system* is a smooth homeomorphism $\psi \colon U \to \mathbb{R}^n$.

Theorem 522. If ψ is a local coordinate system, then

$$\left\{ e_\mu \;\middle|\; e_\mu(f) = \frac{\partial}{\partial x^\mu} f \circ \psi^{-1}|_p \right\}$$

is a basis for TM_p.

Definition 238. Let M be a smooth manifold. The *tangent bundle* is the manifold

$$TM \equiv \{(p, v) \mid p \in M, v \in TM_p\}$$

Theorem 523. $\dim TM = 2 \dim M$

Definition 239. A *vector field* is a smooth map $v \colon M \to TM$ such that $\forall p \in M,\ v(p)_1 = p$.

Definition 240. A *curve* in M is a smooth map $c \colon \mathbb{R} \to M$.

Definition 241. Let c be a curve. The *tangent* to c at $p = c(x_0)$ is $t \in TM_p$ such that

$$t(f) = \frac{d}{dx} f \circ c|_{x_0}$$

Definition 242. A *1-parameter* group is a topological group G which is also a 1-dimensional smooth manifold with a smooth homomorphism $\phi\colon \mathbb{R} \to G$.

Theorem 524. Let ϕ be a 1-parameter group of diffeomorphisms of M. Then the tangents to the curves $\phi(p)$ give a vector field on M.

Theorem 525. For any vector field v on M, there is a unique curve passing through $p \in M$ having v as its tangents. It is called the *integral curve* of v through p.

Theorem 526. The integral curves of a vector field constitute a 1-parameter group of diffeomorphisms.

Definition 243. Let v, w be vector fields. Their *commutator* is

$$[v, w] \equiv v \circ w - w \circ v$$

Theorem 527. $[v, w]$ is a vector field.

Definition 244. Let $\phi\colon M \to N$ be smooth. Let $f \in \mathcal{F}_N$, $v \in TM_p$, $\omega \in TN^*_{\phi(p)}$.
Define the *pullback* $\phi_\star f \in \mathcal{F}_M$ such that $\phi_\star f = f \circ \phi$.
Define the *pushforward* $\phi^\star v \in TN_{\phi(p)}$ such that $\phi^\star v(f) = v(\phi_\star f)$.
Define the *pullback* $\phi_\star \omega \in TM^*_p$ such that $\phi_\star \omega(v) = \omega(\phi^\star v)$.

Definition 245. Let $F\colon M \to N$ be smooth. The *derivative* of F at $p \in M$ is the linear map $DF_p\colon TM_p \to TN_{F(p)}$ such that $\forall f \in \mathcal{F}_N$,

$$DF_p(v)(f) = v(f \circ F)$$

Theorem 528. $D(F \circ G) = DF \cdot DG$

Theorem 529. $T(M \times N)_{p,q} = TM_p \oplus TN_q$, $D(F \times G) = DF \oplus DG$

Theorem 530. Let ψ be a local coordinate system. $D\psi^\mu{}_a = e^\mu_a$.

Theorem 531. If ϕ is a diffeomorphism and T is an arbitrary tensor field, then its pushforward is

$$\phi^\star T = D\phi^{a_1}{}_{b_1} \cdots D\phi^{a_l}{}_{b_l} \, T^{b_1 \ldots b_l}{}_{c_1 \ldots c_k} \, D\phi^{-1}{}^{c_1}{}_{d_1} \cdots D\phi^{-1}{}^{c_k}{}_{d_k}$$

Facts:

$\phi^\star(\alpha T + \beta S) = \alpha \phi^\star T + \beta \phi^\star S$
$\phi^\star(T \otimes S) = \phi^\star T \otimes \phi^\star S$
$\phi^\star(T^a{}_a) = (\phi^\star T)^a{}_a$
$\phi^{-1\star} = \phi^{\star-1} = \phi_\star$

Definition 246. Let v be a vector field and ϕ the corresponding 1-parameter group of diffeomorphisms. Let T be a tensor field. The *Lie derivative* of T with respect to v is

$$\pounds_v T \equiv \frac{d}{dt} \phi^\star(-t) T|_{t=0}$$

Facts:

$\pounds_v(\alpha T + \beta S) = \alpha \pounds_v T + \beta \pounds_v S$
$\pounds_v(T \otimes S) = \pounds_v T \otimes S + T \otimes \pounds_v S$
$\pounds_v(T^a{}_a) = (\pounds_v T)^a{}_a$
$\pounds_v \delta^a{}_b = 0$
$\pounds_v w = [v, w]$
$\pounds_v \pounds_w - \pounds_w \pounds_v = \pounds_{[v,w]}$
$\pounds_{\alpha v + \beta w} = \alpha \pounds_v + \beta \pounds_w$
$\pounds_v T = 0$ iff $\phi^\star T = T$

Definition 247. A function F is a *local diffeomorphism* at p iff p has an open neighborhood U such that F is diffeomorphic on U.

Theorem 532. F is a local diffeomorphism at p iff $\det DF_p \neq 0$.

Definition 248. F is an *immersion* at p iff DF_p is injective.

Definition 249. F is a *submersion* at p iff DF_p is surjective.

Theorem 533. If F, G are immersions, then so are $F \times G$ and $F \circ G$.

Definition 250. Let $F\colon M \to N$. $p \in M$ is *critical* iff DF_p is not surjective. $q \in N$ is *regular* iff DF_p is surjective $\forall p \in F^{-1}(q)$.

Theorem 534. (Sard) Let $F\colon M \to N$ be smooth. Let μ be a Borel measure on N quasi-invariant under diffeomorphisms. Then the set of critical values of F has measure zero in μ.

Theorem 535. If q is a regular value of $F\colon M \to N$, then $F^{-1}(q)$ is a submanifold of M, and $\dim F^{-1}(q) = \dim M - \dim N$.

Theorem 536. Let $F\colon M \to N$ be smooth, let $q \in N$ be regular, let $K \equiv F^{-1}(q)$. Then $\forall p \in K$, $\ker DF_p = TK_p$.

Definition 251. $\{f_i\} \subseteq \mathcal{F}$ are *independent* iff $\{Df_i\}$ are linearly independent.

Theorem 537. If $\{f_i\}$, $i = 1...k$ are independent then $\bigcap_i g_i^{-1}(x)$ is a submanifold of dimension $\dim M - k$.

Definition 252. Let $F\colon M \to N$ be smooth, let K be a submanifold of N. F is *transverse* to K iff $\forall p \in F^{-1}(K)$, $DF_p(TM_p) + TK_{F(p)} = TN_{F(p)}$.

Theorem 538. If F is transverse to K, then $F^{-1}(K)$ is a submanifold, and $\dim F^{-1}(K) = \dim K + \dim M - \dim N$.

Definition 253. Let $F, G\colon M \to N$ be smooth. A *homotopy* from f to g is a smooth function $F\colon M \times [0,1] \to N$ such that $F(p, 0) = f(p)$ and $F(p, 1) = g(p)$. f is *homotopic* to g iff there exists a homotopy between them.

Theorem 539. Homotopy is an equivalence relation on smooth functions.

Definition 254. A manifold M is *contractible* iff the identity map $I\colon M \to M$ is homotopic to a constant.

Theorem 540. A manifold M is contractible iff all maps into M are homotopic.

Theorem 541. Every contractible manifold is connected.

Theorem 542. No manifold is both contractible and compact.

Definition 255. A *derivative operator* is a map ∇ from tensor fields of type (l, k) to tensor fields of type $(l, k + 1)$ satisfying
1. $\nabla(\alpha T + \beta S) = \alpha \nabla T + \beta \nabla S$
2. $\nabla(T \otimes S) = T \otimes \nabla S + \nabla T \otimes S$
3. $\nabla_a \delta^b{}_c = 0$
4. $\forall f \in \mathcal{F}, \ \nabla f = Df$
5. $\forall f \in \mathcal{F}, \ \nabla_a \nabla_b f = \nabla_b \nabla_a f$

Theorem 543. $\nabla(T^a{}_a) = (\nabla T)^a{}_a$

Definition 256. Let ψ be a local coordinate system with basis $\{e_\mu\}$. The *coordinate derivative* is the derivative operator ∂ such that in basis components,
$$\partial_\mu T^{\nu\cdots}{}_{\lambda\cdots} = \frac{\partial}{\partial x^\mu} T^{\nu\cdots}{}_{\lambda\cdots}$$

Theorem 544. Given any two derivative operators $\nabla, \tilde{\nabla}$ there is a unique tensor field $\Gamma^a{}_{bc}$ such that for all vector fields v^a,
$$\nabla_a v^b = \tilde{\nabla}_a v^b + \Gamma^b{}_{ac} v^c$$

Theorem 545. For any tensor field T,
$$\nabla_a T^{b\cdots}{}_{c\cdots} = \tilde{\nabla}_a T^{b\cdots}{}_{c\cdots} + \sum \Gamma^b{}_{ad} T^{d\cdots}{}_{c\cdots} - \sum \Gamma^d{}_{ac} T^{b\cdots}{}_{d\cdots}$$

Theorem 546. $\Gamma^a{}_{bc} = \Gamma^a{}_{cb}$

Definition 257. A curve is a *geodesic* with respect to a derivative operator ∇ iff its tangent satisfies
$$t^a \nabla_a t^b = 0$$

Definition 258. Let $\phi \colon M \to N$ be smooth. Let ∇ be a derivative operator on N. Define the pullback $\phi_\star \nabla$ as the derivative operator on M such that for all dual vector fields ω on N,
$$\phi_\star \nabla(\phi_\star \omega) = \phi_\star(\nabla \omega)$$

Theorem 547. Let ψ be a local coordinate system. Let D denote the derivative on $\mathbb{R}^n$. The coordinate derivative $\partial = \psi_\star D$.

Theorem 548. Let ∇ be any derivative operator. The Lie derivative is given by

$$\pounds_v T = v^c \nabla_c T^{a\cdots}{}_{b\ldots} - \sum T^{c\cdots}{}_{b\ldots} \nabla_c v^a + \sum T^{a\cdots}{}_{c\ldots} \nabla_b v^c$$

Theorem 549. Let M be connected and $f \in \mathcal{F}_m$. $\nabla f = 0$ iff f is constant.

Theorem 550. Let m be the number of connected components of M. On $\mathcal{F}_M$, $\dim \ker \nabla = m$.

Definition 259. The *curvature* of a derivative operator is the unique tensor field $R_{abc}{}^d$ such that for all vector fields v^a,

$$(\nabla_a \nabla_b - \nabla_b \nabla_a) v^c = -R_{abd}{}^c v^d$$

Theorem 551. For any tensor field T,

$$(\nabla_a \nabla_b - \nabla_b \nabla_a) T^{c\cdots}{}_{d\ldots} = -\sum R_{abe}{}^c T^{e\cdots}{}_{d\ldots} + \sum R_{abd}{}^e T^{c\cdots}{}_{e\ldots}$$

Theorem 552. $R_{[abc]}{}^d = R_{(ab)c}{}^d = 0$

Theorem 553. (Bianchi) $\nabla_{[a} R_{bc]d}{}^e = 0$

Theorem 554. Any coordinate derivative ∂ satisfies $R_{abc}{}^d = 0$.

Definition 260. A *p-form* is a totally antisymmetric tensor field of type $(0, p)$. Let Ω_p denote the vector space of p-forms on a manifold.

Definition 261. Define a linear map $d \colon \Omega_p \to \Omega_{p+1}$. For any derivative operator ∇,

$$d\omega \equiv (p+1) \nabla_{[a} \omega_{bc\ldots]}$$

Theorem 555. d is independent of derivative operator.

Theorem 556. $d^2 = 0$

Theorem 557. $d(\phi_\star \omega) = \phi_\star(d\omega)$

Theorem 558. (Cartan) On p-forms, let i_v denote contraction with the vector field v in the first index.
$$\pounds_v = i_v d + d i_v$$
$$\pounds_v i_w - i_w \pounds_v = i_{[v,w]}$$
$$\pounds_v d = d \pounds_v$$

Theorem 559. Let $\dim M = n$, let ϵ be an n-form. Let ψ be a local coordinate system. $\epsilon(e_\mu) = \det D\psi^{-1}$.

Theorem 560. Let $\dim M = n$, let ϵ be an n-form. For $U \subseteq M$, let $\psi \colon U \to \mathbb{R}^n$ be a local coordinate system. Then $\int_{\psi(U)} |\epsilon(e_\mu)|$ is independent of ψ and

$$\mu_\epsilon(U) \equiv \int_{\psi(U)} |\epsilon(e_\mu)|$$

is a Borel measure on M.

Theorem 561. $\mu_{\alpha\epsilon + \beta\varepsilon} = \alpha\mu_\epsilon + \beta\mu_\varepsilon$

Theorem 562. Let $f \in \mathcal{F}$.

$$\frac{d\mu_{f\epsilon}}{d\mu_\epsilon} = |f|$$

Theorem 563. Let ϕ be a diffeomorphism. $\phi(\mu_\epsilon) = \mu_{\phi^\star\epsilon}$ and

$$\frac{d\phi(\mu_\epsilon)}{d\mu_\epsilon} = |\det D\phi^{-1}|$$

Theorem 564. Let ϕ be a diffeomorphism.

$$\int_{\phi(U)} f = \int_U f \circ \phi \, |\det D\phi|$$

Definition 262. A *manifold with boundary* is a 2nd countable Hausdorff space M such that every point has a neighborhood homeomorphic to an open subset of $\{x \in \mathbb{R}^n \mid x_1 \geq 0\}$. Its *boundary* ∂M is the subset of M consisting of points with no neighborhood homeomorphic to an open subset of $\mathbb{R}^n$.

Theorem 565. Let $\dim M = n$. ∂M is a manifold, $\dim \partial M = n - 1$, and $\partial \partial M = \varnothing$.

Theorem 566. Let $\partial N = \varnothing$. Then $\partial(M \times N) = \partial M \times N$.

Theorem 567. (Stokes) Let $\dim M = n$, let ω be an $n - 1$ form.

$$\int_M d\omega = \int_{\partial M} \omega$$

Definition 263. A *Riemannian* manifold is a smooth manifold with a tensor field g_{ab} such that g_{ab} is an inner product on each tangent space.

Theorem 568. Every submanifold of a Riemannian manifold is Riemannian, under the pullback of g_{ab} by the identity map.

Theorem 569. On a Riemannian manifold, there is a unique derivative operator ∇ satisfying $\nabla_a g_{bc} = 0$.

Theorem 570. On a Riemannian manifold, let $\tilde{\nabla}$ be any derivative operator. The derivative operator ∇ associated with g_{ab}, and its curvature, are given by

$$\Gamma^a{}_{bc} = \frac{1}{2} g^{ad} (\tilde{\nabla}_c g_{db} + \tilde{\nabla}_b g_{dc} - \tilde{\nabla}_d g_{bc})$$

$$R_{abc}{}^d = -2\tilde{\nabla}_{[a} \Gamma^d{}_{b]c} + 2\Gamma^e{}_{c[a} \Gamma^d{}_{b]e}$$

Theorem 571. For the derivative operator associated with g_{ab}, $R_{ab(cd)} = 0$.

Definition 264. $R_{ab} \equiv R_{acb}{}^c$, $R \equiv R^a{}_a$

Theorem 572. $R_{ab} = R_{ba}$

Theorem 573. $\nabla_a (R^a{}_b - \frac{1}{2} R \delta^a{}_b) = 0$

Definition 265. A vector field v on a Riemannian manifold is a *Killing field* iff $\nabla_a v_b + \nabla_b v_a = 0$.

Theorem 574. $\mathcal{L}_v g_{ab} = \nabla_a v_b + \nabla_b v_a$, so the 1-parameter group of diffeomorphisms generated by v preserves g_{ab} iff v is Killing.

Theorem 575. If v and w are Killing, then so are $\alpha v + \beta w$ and $[v, w]$.

Theorem 576. Let v be a Killing field and c be a geodesic with tangent t. Then $v_a t^a$ is constant along c.

Theorem 577. On a Riemannian manifold of dimension n, there is a unique n-form ϵ and measure μ_ϵ such that $\langle \epsilon, \epsilon \rangle = n!$.

Theorem 578. $\nabla \epsilon = 0$

Theorem 579. (Fundamental theorem of calculus) Let V be a finite-dimensional vector space with inner product. Let X be a Banach space with basis. Let M be a submanifold of V with boundary, with outward orthonormal vector n. Let $f \colon M \to X$ be smooth. Then

$$\int_M Df = \int_{\partial M} f \otimes n^\dagger$$

Theorem 580. (Cauchy) Let $f \colon \mathbb{C} \to \mathbb{C}$ be differentiable.

Let $c \colon [0, 1] \to \mathbb{C}$ be a curve in $\mathbb{C}$. Then $\int_c f$ depends only on $c(0)$ and $c(1)$.

$$\int_c \frac{df}{dz} = f(c(1)) - f(c(0))$$

$$\frac{d}{dz} \int_{z_0}^z f = f$$

Let U be a 2-dimensional submanifold of $\mathbb{C}$ with boundary. Let $\{a_i\} \subset U$.

$$\int_{\partial U} f = 0$$

$$\int_{\partial U} \frac{f}{\prod_i (z - a_i)^{n_i + 1}} = 2\pi i \sum_i \frac{D^{n_i} f(a_i)}{n_i!}$$

Definition 266. Let v be a vector field on a Riemannian manifold. The *divergence* of v is the field $\nabla \cdot v \in \mathcal{F}$ such that

$$\pounds_v \epsilon = (\nabla \cdot v)\epsilon$$

Definition 267. On a Riemannian manifold, the *Hodge star* is the linear isomorphism $*\colon \Omega_p \to \Omega_{n-p}$ given by

$$*\omega_{a_1 \ldots a_{n-p}} \equiv \frac{1}{p!}\omega^{b_1 \ldots b_p} \epsilon_{b_1 \ldots b_p a_1 \ldots a_{n-p}}$$

Theorem 581. $**\omega = (-1)^{p(n-p)}\omega$

Theorem 582. $\langle *\alpha, *\beta \rangle = \frac{(n-p)!}{p!}\langle \alpha, \beta \rangle$

Definition 268. The *codifferential* is the linear map $\delta\colon \Omega_p \to \Omega_{p-1}$ given by

$$\delta \equiv (-1)^{n(p+1)} * d *$$

Theorem 583. $\delta^2 = 0$

Theorem 584. Let $\langle\langle , \rangle\rangle$ denote the natural inner product of p-forms in $L_2(\mu_\epsilon)$.

$$\langle\langle \alpha, \delta\beta \rangle\rangle = -\langle\langle d\alpha, \beta \rangle\rangle$$

Theorem 585. $\delta\omega = \nabla^a \omega_{ab\ldots}$

Theorem 586. $\delta v = \nabla \cdot v$

Definition 269. The *Laplacian* is the linear map $\nabla^2 \equiv d\delta + \delta d$.

Theorem 587. $\forall \omega, \ \langle\langle \omega, \nabla^2 \omega \rangle\rangle \leq 0$

Facts:

$\nabla^2 * = *\nabla^2$
$\nabla^2 d = d\nabla^2$
$\nabla^2 \delta = \delta\nabla^2$

Theorem 588. Let M be compact, connected. Let $f \in \mathcal{F}_M$. If $\nabla^2 f = 0$ then f is constant.

Theorem 589. Let M be compact with boundary. The eigenfunctions of ∇^2 are an orthonormal basis for $L_2(\mu_\epsilon)$.

Theorem 590. (Hodge) Let M be compact, let $H \equiv \ker \nabla^2$.
$H = \ker d \cap \ker \delta$
$\Omega_p = \nabla^2 \Omega_p \oplus H$ orthogonally
$\nabla^2 \Omega_p = d\Omega_{p-1} \oplus \delta\Omega_{p+1}$ orthogonally
$\ker d = d\Omega_{p-1} \oplus H$ orthogonally
$\ker \delta = \delta\Omega_{p+1} \oplus H$ orthogonally

Theorem 591. (Poincare) If M is contractible then $\ker d = d\Omega_{p-1}$ and $\ker \delta = \delta\Omega_{p+1}$.

Theorem 592. If M is contractible, then the Laplacian is surjective: $\forall \alpha \in \Omega_p$, $\exists \beta \in \Omega_p$ such that $\nabla^2 \beta = \alpha$.

Theorem 593. If M is contractible, then $\Omega_p = d\Omega_{p-1} + \delta\Omega_{p+1}$.

Theorem 594. Let M be contractible. If $d\alpha = 0$ and $\delta\beta = 0$, then $\exists \omega$ such that $d\omega = \alpha$ and $\delta\omega = \beta$.

4.4   Lie Theory

Definition 270. A *Lie algebra* is a vector space $\mathcal{L}$ with a bilinear map $[,] \colon \mathcal{L} \times \mathcal{L} \to \mathcal{L}$ satisfying $\forall v, u, w \in \mathcal{L}$,
1. (antisymmetric) $[v, u] = -[v, u]$
2. (Jacobi identity) $[v, [u, w]] + [u, [w, v]] + [w, [v, u]] = 0$

Definition 271. Let $\mathcal{L}$ be a Lie algebra. The *structure constants* are the unique tensor $c^a{}_{bc}$ such that $\forall v, w \in \mathcal{L}$,

$$[v, w]^a = c^a{}_{bc} v^b w^c$$

Theorem 595. Let $\{T_\mu\}$ be a basis for $\mathcal{L}$. Then

$$[T_\mu, T_\nu] = c^\alpha{}_{\mu\nu} T_\alpha$$

Definition 272. A *Lie group* is a group which is also a manifold such that group composition and inverse are smooth.

Theorem 596. Let V be a finite-dimensional vector space over $\mathbb{R}$ or $\mathbb{C}$. $GL(V)$, $SL(V)$, and $U(V, g)$ are Lie groups. $U(V, g)$ is compact.

Theorem 597. Let G be a Lie group, let e be the identity, and v_e a tangent vector at e. Then there is a unique vector field v on G such that $v(e) = v_e$, and which is invariant under the left action of the group: $\forall g \in G$, $\forall \phi \in \mathcal{F}$,

$$v(g)(\phi) = v_e(\phi \circ g)$$

Theorem 598. Let G be an n-dimensional Lie group. The set of G-invariant vector fields on G is an n-dimensional Lie algebra over $\mathbb{R}$, with the bracket given by the commutator of vector fields. It is denoted $\pounds G$.

Theorem 599. Let H be a 1-parameter subgroup of G. Then the right cosets Hg are a family of smooth curves, with tangent vectors giving a G-invariant vector field $v_H \in \pounds G$.

Theorem 600. (exponential map) Let $v \in \pounds G$. The integral curve of v through e is a 1-parameter subgroup of G.

Theorem 601. Let H, K be 1-parameter subgroups of G. $[v_H, v_K] = 0$ iff $hk = kh \; \forall h \in H \; \forall k \in K$.

Theorem 602. Let $f \colon G \to H$ be a smooth homomorphism. Then its derivative $Df \colon \pounds G \to \pounds H$ is a linear map satisfying $[Df(v), Df(w)] = Df([v, w])$, and $\pounds \ker f = \ker Df$.

Theorem 603. $H \leq G$ iff $\pounds H \leq \pounds G$

Theorem 604. Let $H, K \leq G$. Then
$\pounds(H \cap K) = \pounds H \cap \pounds K$
$\pounds(H \times K) = \pounds H \times \pounds K$

Theorem 605. $£GL(V) = \mathcal{L}(V)$ with the bracket given by $[A, B] = AB - BA$.
$£SL(V) = \{L \mid \mathrm{Tr}L = 0\}$
$£U(V, g) = \{L \mid L^\dagger = -L\}$

Theorem 606. Let $G \leq GL(V)$ be connected. $L \in £G$ iff $\exp(L) \in G$. $\exp(tL)$ is the 1-parameter subgroup integral curve of L.

Theorem 607. Let A be an algebra over $\mathbb{R}$. $£\mathrm{Aut}(A) = \{D \in \mathcal{L}(A) \mid D(ab) = D(a)b + aD(b)\}$.

Theorem 608. The set of vector fields on a manifold is a Lie algebra under commutator. It is associated to Lie groups of diffeomorphisms.

Definition 273. Let $£$ be a Lie algebra. The *adjoint representation* is the linear map ad: $£ \to \mathcal{L}(£)$ given by

$$\mathrm{ad}(v)(w) \equiv [v, w]$$

In index notation, $\mathrm{ad}(v)^a{}_b = c^a{}_{cb}v^c$.

Theorem 609. $[\mathrm{ad}(v), \mathrm{ad}(w)] = \mathrm{ad}([v, w])$

Definition 274. The *Killing form* on $£$ is the natural bilinear form $K \colon £ \times £ \to \mathbb{R}$ given by

$$K(v, w) \equiv -\mathrm{Tr}(\mathrm{ad}(v)\mathrm{ad}(w))$$

In index notation, $K_{ij} = -c^a{}_{ib}c^b{}_{ja}$.

Facts:

$K(v, w) = K(w, v)$
$K([v, w], u) = K(v, [w, u])$

Definition 275. A Lie algebra is *simple* iff it is non-abelian with no proper, nontrivial ideals.

Definition 276. A Lie algebra is *semisimple* iff it has no nontrivial abelian ideals.

Theorem 610. A Lie algebra is semisimple iff it is a direct sum of simple Lie algebras.

Theorem 611. If $\mathcal{L}$ is semisimple then its adjoint representation is injective.

Theorem 612. A Lie algebra is semisimple iff its Killing form is invertible.

Theorem 613. If G is compact, then the Killing form on $\mathcal{L}G$ is positive.

Theorem 614. K is an inner product on $\mathcal{L}G$ iff G is compact and $\mathcal{L}G$ is semisimple.

Theorem 615. Let $\mathcal{L}G$ be compact, semisimple. $\forall v \in \mathcal{L}G$, $\mathrm{ad}(v)^\dagger = -\mathrm{ad}(v)$ with respect to K.

Theorem 616. Let $\mathcal{L}G$ be compact, semisimple. If $V, W \leq \mathcal{L}G$ are ideals and $V \cap W = \{0\}$, then V and W are orthogonal.

Theorem 617. Let $\mathcal{L}G$ be compact, semisimple. If $V \leq \mathcal{L}G$ is an ideal, then so is $V^\perp$.

Theorem 618. Let $\dim V = n$. In $\mathcal{L}GL(V)$,

$$K(A, B) = 2\mathrm{Tr}A\mathrm{Tr}B - 2n\mathrm{Tr}(AB)$$

4.5 Representation Theory

Definition 277. Let G be a topological group, let V be a topological vector space. A *representation* of G on V is a continuous homomorphism $R\colon G \to GL(V)$. It is *faithful* iff it is injective.

Definition 278. Two representations R_1, R_2 of G on V are *equivalent* iff $\exists U \in GL(V)$ such that $\forall g \in G$, $U R_1(g) U^{-1} = R_2(g)$.

Theorem 619. On any set of representations, equivalence is an equivalence relation.

Theorem 620. Let R be a representation of G on V. If there exists a proper, nontrivial subspace $W < V$ such that $R(G)(W) \subseteq W$, then the restriction of R to W is a representation of G.

Definition 279. A representation R of G on V is *irreducible* iff there is no proper, nontrivial subspace $W < V$ such that $R(G)(W) \subseteq W$.

Definition 280. Let R_1, R_2 be representations of G on V_1, V_2 respectively. Their *direct sum* is the representation $R_1 \oplus R_2$ on $V_1 \oplus V_2$ given by

$$(R_1 \oplus R_2)(g)(v_1 \oplus v_2) \equiv R_1(g)(v_1) \oplus R_2(g)(v_2)$$

Definition 281. A representation is *completely reducible* iff it is equivalent to a direct sum of irreducible representations.

Theorem 621. Let G be compact. Let V be finite dimensional. Every representation R of G on V is orthogonal/unitary in the inner product

$$\langle v, w \rangle' \equiv \frac{1}{\mu(G)} \int_G \langle R(g)v, R(g)w \rangle \ d\mu$$

where $\langle, \rangle$ is any inner product, and μ is the Haar measure.

Theorem 622. Every finite-dimensional orthogonal/unitary representation is completely reducible, and the irreducible subspaces are orthogonal.

Theorem 623. Let G be compact. Every complex unitary representation of G is completely reducible, and the irreducible subspaces are orthogonal.

Theorem 624. Let G be compact. Every complex irreducible representation of G is finite-dimensional.

Theorem 625. Every locally compact group has a unitary representation given by the *regular* representation λ on $L_2(G, \mathbb{C})$ with Haar measure:

$$\lambda(g)f(h) \equiv f(g^{-1}h)$$

Theorem 626. The regular representation is faithful.

Theorem 627. (Schur) Let R_1, R_2 be irreducible complex representations of G on V, W respectively. Let $L \in \mathcal{L}(V, W)$ such that $\forall g \in G$,

$$LR_1(g) = R_2(g)L$$

then the following cases hold:
1. If R_1, R_2 are not equivalent, then $L = 0$.
2. If R_1, R_2 are equivalent, then L is invertible.
3. If $R_1 = R_2$, then $L = \lambda \mathbf{1}$ for some $\lambda \in \mathbb{C}$.

If L, L' are two such maps, then $L' = \lambda L$ for some $\lambda \in \mathbb{C}$.

Theorem 628. If R is a representation of G and $n \in \mathbb{Z}$ then $(\det R(g))^n$ gives a one-dimensional representation of G.

Theorem 629. If R is a representation of G on V, then $R^T(g) \equiv (R(g^{-1}))^T$ is a representation of G on V^*.

Theorem 630. R^T is irreducible iff R is.

Definition 282. Let R_1, R_2 be representations of G on V_1, V_2 respectively. Their *tensor product* is the representation $R_1 \otimes R_2$ on $V_1 \otimes V_2$ given by

$$(R_1 \otimes R_2)(g)(v_1 \otimes v_2) \equiv R_1(g)(v_1) \otimes R_2(g)(v_2)$$

Definition 283. The *characters* of a representation R are the function $\chi_R(g) \equiv \mathrm{Tr} R(g)$.

Theorem 631. The characters are constant on the conjugacy classes of G.

Theorem 632.

$$\chi_{R_1 \oplus R_2}(g) = \chi_{R_1}(g) + \chi_{R_2}(g)$$

$$\chi_{R_1 \otimes R_2}(g) = \chi_{R_1}(g)\chi_{R_2}(g)$$

Theorem 633. Let G be compact. Indexing the complex irreducible representations with Greek letters, let representation R_α have dimension n_α. Then

$$\int_G R_\alpha(g^{-1})^a{}_b R_\beta(g)^c{}_d \; d\mu = \frac{\mu(G)}{n_\alpha} \delta_{\alpha\beta} \delta^a{}_d \delta^c{}_b$$

Theorem 634. Let G be compact. Indexing the complex irreducible representations with Greek letters,

$$\int_G \overline{\chi_\alpha(g)} \chi_\beta(g) \; d\mu = \mu(G) \delta_{\alpha\beta}$$

Theorem 635. Let G be compact, let R be a complex unitary representation of G. By complete reducibility, R is a direct sum of irreducibles:

$$R = \bigoplus m_\alpha R_\alpha$$

where the multiplicity m_α is given by

$$m_\alpha = \frac{1}{\mu(G)} \int_G \overline{\chi_\alpha(g)} \chi_R(g) \; d\mu$$

Theorem 636. Let G be compact, let R be a complex unitary representation of G. Define

$$P_\alpha \equiv \frac{n_\alpha}{\mu(G)} \int_G \overline{\chi_\alpha(g)} R(g) \; d\mu$$

Then $P_\alpha^2 = P_\alpha$, $P_\alpha^\dagger = P_\alpha$, and P_α projects onto the subspace transforming under R_α.

Theorem 637. Let G be compact. For complex irreducible representations $R_\alpha, R_\beta, R_\gamma$, the tensor product $R_\alpha \otimes R_\beta = \bigoplus N_{\alpha\beta\gamma} R_\gamma$, where

$$N_{\alpha\beta\gamma} = \frac{1}{\mu(G)} \int_G \overline{\chi_\alpha(g)\chi_\beta(g)} \chi_\gamma(g) \; d\mu$$

Theorem 638. Let G be compact, let R be a complex unitary representation of G. R is irreducible iff

$$\frac{1}{\mu(G)} \int_G \overline{\chi_R(g)} \chi_R(g) \, d\mu = 1$$

Theorem 639. Let G be compact, let R, S be complex unitary representations of G. R is equivalent to S iff $\chi_R = \chi_S$.

Theorem 640. Let G, H be compact, let R be a complex unitary representation of $G \times H$. R is irreducible iff $R = R_1 \otimes R_2$, where R_1, R_2 are complex irreducible representations of G, H respectively.

Theorem 641. If G is finite, then the number of complex irreducible representations of G equals the number of conjugacy classes in G.

Theorem 642. Let G be finite. Index the conjugacy classes $C_i \subseteq G$. Index the complex irreducible representations R_α. Then

$$\sum_\alpha \overline{\chi_\alpha(C_i)} \chi_\alpha(C_j) = \frac{\#G}{\#C_i} \delta_{ij}$$

Theorem 643. Let G be finite and let n_α be the dimensions of the complex irreducible representations of G.

$$\sum_\alpha n_\alpha^2 = \#G$$

Theorem 644. Let G be finite and let n_α be the dimensions of the complex irreducible representations of G. $\forall \alpha$, $\exists m \in \mathbb{N}$ such that $\#G = mn_\alpha$.

Definition 284. Let G be locally compact with Haar measure μ. *Convolution* is the bilinear map $*: L_p(\mu, \mathbb{C}) \times L_q(\mu, \mathbb{C}) \to L_r(\mu, \mathbb{C})$, where $\frac{1}{p} + \frac{1}{q} = \frac{1}{r} + 1$, given by

$$(f * k)(h) \equiv \int_G f(g) k(g^{-1}h) \, d\mu$$

Theorem 645. $L_1(\mu, \mathbb{C})$ is an algebra under $*$, which is commutative iff G is abelian.

Definition 285. Let G be locally compact with Haar measure μ. *Parity* is the linear map $P \colon L_p(\mu, \mathbb{C}) \to L_p(\mu, \mathbb{C})$ given by

$$Pf(g) \equiv f(g^{-1})$$

Facts:

$P(f \cdot k) = Pf \cdot Pk$
$\lambda(g)(f \cdot k) = \lambda(g)f \cdot \lambda(g)k$
$P(f * k) = Pk * Pf$
$\lambda(g)(f * k) = \lambda(g)(f) * k$
$f * 1 = \int f$
$\int f * k = \int f \cdot \int k$

Theorem 646. If G is abelian, then $P\lambda(g) = \lambda(g^{-1})P$

Theorem 647. In $L_2(\mu, \mathbb{C})$,

$$\langle h, f * k \rangle = \langle P\bar{f} * h, k \rangle$$

Theorem 648. If G is abelian, compact, or discrete, then $\forall f$, $\int f = \int Pf$.

Theorem 649. (Peter-Weyl) Let G be compact with Haar measure μ. Indexing the complex irreducible representations R_α acting on V_α with dimensions n_α, then

$$L_2(\mu, \mathbb{C}) \simeq \overline{\bigoplus \mathcal{L}(V_\alpha)}$$

under the linear isomorphism

$$\mathcal{F}(f) \equiv \bigoplus \int_G f(g) R_\alpha(g) \, d\mu$$

and

$$\mathcal{F}^{-1}\left(\bigoplus L_\alpha\right)(g) = \sum n_\alpha \mathrm{Tr}(R_\alpha(g)^\dagger L_\alpha)$$

Theorem 650. $\mathcal{F}$ is unitary: on $\overline{\bigoplus \mathcal{L}(V_\alpha)}$, define the inner product

$$\langle L, T \rangle \equiv \mu(G) \sum n_\alpha \operatorname{Tr}(L_\alpha^\dagger T_\alpha)$$

then

$$\langle \mathcal{F}(f), \mathcal{F}(k) \rangle = \langle f, k \rangle$$

Theorem 651. $\mathcal{F}$ is an isomorphism of convolution:

$$\mathcal{F}(f * k) = \mathcal{F}f \cdot \mathcal{F}k$$

Theorem 652. $(\mathcal{F}f)^\dagger = \mathcal{F}P\bar{f}$

Theorem 653. $\mathcal{F}$ is an isomorphism of the regular representation:

$$\mathcal{F}(\lambda(g)f) = \bigoplus R_\alpha(g) \cdot \mathcal{F}f$$

Theorem 654. The regular representation decomposes in irreducibles as $\lambda = \bigoplus n_\alpha R_\alpha$.

Theorem 655. All complex irreducible representations of an abelian group are 1-dimensional.

Definition 286. Let G be locally compact abelian. Its *dual group* is the group $\hat{G}$ of the unitary irreducible representations of G, with the group product $\otimes$.

Theorem 656. $\hat{G}$ is also locally compact abelian.

Theorem 657. Let G be locally compact abelian. The elements of $\hat{G}$ are linearly independent in $L_\infty(G, \mathbb{C})$. If G is compact, then $\hat{G}$ is an orthonormal basis for $L_2(\mu, \mathbb{C})$.

Theorem 658. $\widehat{G \times H} \simeq \hat{G} \times \hat{H}$

Theorem 659. If G is finite, then $G \simeq \hat{G}$.

Theorem 660. (Pontryagin) $\hat{\hat{G}} = G$ under the identification $g(\chi) \equiv \chi(g)$ for $g \in G$, $\chi \in \hat{G}$.

Theorem 661. $\hat{\mathbb{R}} = \mathbb{R}$, $\hat{\mathbb{Z}} = U(1)$

Theorem 662. G is compact iff $\hat{G}$ is discrete.

Definition 287. Let G be locally compact abelian with Haar measure μ. The *Fourier transform* is the linear map $\mathcal{F} \colon L_2(G, \mathbb{C}) \to L_2(\hat{G}, \mathbb{C})$ given by

$$\mathcal{F}f(\chi) \equiv \int_G \bar{\chi}(g) f(g) \ d\mu$$

Theorem 663. $\mathcal{F}$ is a unitary isomorphism.

$$\langle f, k \rangle = \langle \mathcal{F}f, \mathcal{F}k \rangle$$

Theorem 664. (Fourier)

$$f(g) = \int_{\hat{G}} \mathcal{F}f(\chi) \chi(g) \ d\hat{\mu}$$

Theorem 665.
$$\mathcal{F}(f * k) = \mathcal{F}f \cdot \mathcal{F}k$$
$$\mathcal{F}(f \cdot k) = \mathcal{F}f * \mathcal{F}k$$

Facts:

$\mathcal{F}^2 = P$
$\mathcal{F}P = P\mathcal{F}$
$\overline{\mathcal{F}f} = \mathcal{F}P\bar{f}$
$\mathcal{F}\lambda(g)f(\chi) = \chi(g)\mathcal{F}f(\chi)$
$\lambda(\chi)\mathcal{F}f = \mathcal{F}(\chi \cdot f)$

Glossary of notation

iff	if and only if
$\forall$	for all
$\exists$	there exists
$\Rightarrow$	implies
$\equiv$	defined as
$\in$	is an element of
$\{x \mid P(x)\}$	the set containing x such that $P(x)$
$\varnothing$	empty set
$x \subseteq y$	subset
$x \sim y$	equivalence relation
$A \simeq B$	isomorphism
$\mu \equiv \nu$	equivalence of measures
$x \cup y$	union
$x \cap y$	intersection
$A \times B$	Cartesian product of sets
$f \times g$	Cartesian product of functions
$\mu \times \nu$	product measure
$V \oplus W$	direct sum
$R_1 \oplus R_2$	direct sum representation
$V \otimes W$	tensor product space
$T \otimes S$	tensor product
$R_1 \otimes R_2$	tensor product representation
$A \setminus B$	A without B
$f \circ g$	composition of functions
$f * k$	convolution

$\langle v, w \rangle$	inner product
$a + b$	abelian group operation
$a \cdot b$	ring multiplication
$[v, w]$	commutator of vector fields
$[v, w]$	Lie bracket
$[x, y]$	closed interval
(x, y)	open interval
(x, y)	ordered pair
$K(v, w)$	Killing form
$x < y$	order relation
$H \leq G$	subgroup
$S \leq R$	subring
$E \leq F$	subfield
$W \leq V$	vector subspace
$N \trianglelefteq G$	normal subgroup
$N \trianglelefteq R$	ideal
$E \trianglelefteq F$	normal field extension
$\{x_i\}$	indexed family
$\lim x_n$	limit of a sequence
$\sum_n a_n$	sum
$\prod_i x_i$	Cartesian product of indexed family
$\bigcup X$	union of X
$\bigcup_i x_i$	union of indexed family
$\bigcap X$	intersection of X
$\bigcap_i x_i$	intersection of indexed family
$\bigoplus_i V_i$	direct sum
A^c	compliment
$\mathring{A}$	interior
∂A	boundary
$f^\flat$	average of f under a group action
$\lambda^\sharp$	measure induced by a group action
$v^\dagger$	vector adjoint
$L^\dagger$	linear map adjoint
L^T	transpose

R^T	dual representation
$V^\perp$	orthogonal compliment
$\bar{A}$	topological closure
$\bar{\mu}$	measure completion
$\bar{F}_E$	algebraic closure of F in E
$\bar{F}$	algebraic closure
$\bar{z}$	complex conjugate
$\bar{v}$	vector complex conjugate
$\bar{T}$	tensor complex conjugate
$\bar{V}$	conjugate space
$\bar{V}$	norm completion
V^*	dual space
x^*	*-algebra conjugate
$*\omega$	Hodge star
$\phi_\star$	pullback
$\phi^\star$	pushforward
$\mathrm{char}F$	field characteristic
$\dim_F V$	dimension of V over F
$\mathrm{span}S$	span
$\mathrm{Tr}L$	trace
$\det L$	determinant
$\ker \phi$	kernel
$\mathrm{supp}\mu$	support of a measure
$\sup$	least upper bound
$\inf$	greatest lower bound
$f\colon X \to Y$	function from X to Y
$f(x)$	value of a function
$f(A)$	image of A under f
$f(\mu)$	pushforward of a measure
B^A	set of functions from A to B
$\exp$	exponential function
f^n	exponentiation of f
$f^{-1}(B)$	inverse image of B under f
f^{-1}	inverse function

g^{-1}	group inverse
a^{-1}	multiplicative inverse
$-a$	additive inverse
$T^{a...a'...}{}_{b...b'...}$	tensor
$\mathrm{Sym}T$	symmetric part
$T^{(a...)}$	symmetric part
$\mathrm{Asym}T$	antisymmetric part
$T^{[a...]}$	antisymmetric part
g_{ab}	inner product
$R_{abc}{}^{d}$	curvature tensor
R_{ab}	$R_{acb}{}^{c}$
R	$R^{a}{}_{a}$
ϵ	volume form
$c^{a}{}_{bc}$	structure constants
$\#X$	cardinality
$n!$	factorial
e	group identity
0	additive identity
1	multiplicative identity
$\delta^{i}{}_{j}$	Kronecker delta
$\delta^{a}{}_{b}$	identity linear map
$\lvert x \rvert$	absolute value
$\lVert v \rVert$	norm
$d(x,y)$	metric
P_{V}	projection onto V
Pf	parity
$\mathcal{F}f$	Fourier transform
$\sigma(x)$	spectrum
gH	coset of H
HN	$\{hn \mid h \in H,\ n \in N\}$
$(G:H)$	$\#\{gH \mid g \in G\}$
$\gcd(n,m)$	greatest common divisor
$\mathrm{lcm}(n,m)$	least common multiple
$a\mid b$	a divides b

$b_r(x)$	ball of radius r around x
$\int_E f \, d\mu$	integral of f over E with respect to μ
$\frac{d\mu}{d\nu}$	Radon-Nikodym derivative
Df	derivative of f
∇	derivative operator
∂	coordinate derivative
$d\omega$	exterior derivative
$\delta\omega$	codifferential
$\nabla \cdot v$	divergence
∇^2	Laplacian
$\mathcal{L}_v T$	Lie derivative
$\mathcal{L}G$	Lie algebra of G
$\mathrm{ad}(v)$	adjoint representation
$\lambda(g)$	regular representation
χ_R	characters of R
x^+	successor
$S(n)$	successor
S_X	permutation group of X
$G(x)$	orbit of group action
G_x	$\{g \in G \mid g(x) = x\}$
G_X	$\{g \in G \mid g(x) = x \; \forall x \in X\}$
X_g	$\{x \in X \mid g(x) = x\}$
X_G	$\{x \in X \mid g(x) = x \; \forall g \in G\}$
X/G	$\{G(x) \mid x \in X\}$
$\mathbb{N}$	natural numbers
$\mathbb{Z}$	integers
$\mathbb{Z}_n$	cyclic group order n
$\mathbb{Q}$	rational numbers
$\mathbb{R}$	real numbers
$\mathbb{C}$	complex numbers
$\mathcal{P}(X)$	power set of X
G/N	factor group
$Z(G)$	center of G
$\langle A \rangle$	group generated by A
$\langle a \rangle$	ring $\{ra \mid r \in R\}$

$\mathrm{Aut}(R)$	Automorphism group
$R^\times$	multiplicative group of R
$F(D)$	field of fractions
$P(F)$	polynomial ring
$\mathcal{L}(V, W)$	space of linear maps from V to W
$\mathcal{L}(V)$	$\mathcal{L}(V, V)$
$B(V, W)$	space of continuous linear maps from V to W
$B(V)$	$B(V, V)$
$GL(V)$	general linear group
$SL(V)$	special linear group
$O(V, g)$	orthogonal group
$U(V, g)$	unitary group
$V_T(\lambda)$	eigenspace of $T \in \mathcal{L}(V)$ with eigenvalue λ
$\mathcal{T}(l, k, n, m)$	tensor space
$\beta(X)$	Stone-Čech compactification
$L_p(\mu, V)$	space of p-integrable V-valued functions over measure μ
$L_\infty(X, V)$	space of bounded V-valued functions on X
$C_\infty(V, W)$	space of smooth functions from V to W
$C^*(A)$	enveloping C*-algebra
$\mathrm{Diff}(M)$	diffeomorphism group
$\mathcal{F}_M$	$\{f \colon M \to \mathbb{R} \text{ smooth}\}$
TM_p	tangent space at p
TM	tangent bundle
Ω_p	space of p-forms
$\hat{G}$	dual group

Bibliography

Bourbaki, N. (1989), *Elements of Mathematics*, (Springer-Verlag, New York).

Curtis, M.L. (1990), *Abstract Linear Algebra*, (Springer-Verlag, New York).

Fraleigh, J.B. (2003), *A First Course in Abstract Algebra*, 7th ed., (Pearson Education, Boston).

Gelbaum, B.R., and Olmsted, J.M.H. (1964), *Counterexamples in Analysis*, (Holden-Day, San Francisco).

Halmos, P.R (1974), *Naive Set Theory*, (Springer-Verlag, New York).

Halmos, P.R. (1974), *Measure Theory*, (Springer-Verlag, New York).

Munkres, J.R. (2000), *Topology*, 2nd ed., (Prentice Hall, Upper Saddle River, NJ).

Rudin, W. (1976), *Principles of Mathematical Analysis*, 3rd ed., (McGraw-Hill, New York).

Sally, P.J. (2008), *Tools of the Trade: An Introduction to Advanced Mathematics*, (American Mathematical Society, Providence).

Spivak, M. (2008), *Calculus*, 4th ed., (Publish or Perish, Houston).

Steen, L.A., and Seebach, J.A. (1978), *Counterexamples in Topology*, 2nd ed., (Springer-Verlag, New York).

Index

www.ingramcontent.com/pod-product-compliance
Lightning Source LLC
Chambersburg PA
CBHW080819180526

45168CB00006B/2505